The Silicon Dragon

The Silicon Dragon

High-Tech Industry in Taiwan

Terence Tsai

Associate Director, MBA Programmes, Department of Management, Faculty of Business Administration, The Chinese University of Hong Kong and Senior Research Associate, Judge Business School, University of Cambridge, UK

Bor-Shiuan Cheng

Professor in Organisational Psychology and Chairperson, Department of Psychology, College of Science, National Taiwan University, Taiwan

Edward Elgar

Cheltenham, UK • Northampton, MA, USA

Published by
Edward Elgar Publishing Limited
Glensanda House
Montpellier Parade
Cheltenham
Glos GL50 1UA
UK

Edward Elgar Publishing, Inc.
136 West Street
Suite 202
Northampton
Massachusetts 01060
USA

A catalogue record for this book
is available from the British Library

Library of Congress Cataloguing in Publication Data
Tsai, Terence, 1964–
Silicon dragon : high tech industry in Taiwan / Soo-Hung Terence Tsai, Bor-Shiuan Cheng.
p. cm.
Includes bibliographical references and index.
1. High technology industries—Taiwan. I. Cheng, Bor-Shiuan, 1952– II. Title.
HC430.5.Z9H538 2006
338.4′760951249—dc22 2005032144

ISBN-13: 978 1 84064 240 7
ISBN-10: 1 84064 240 8

Printed and bound in Great Britain by MPG Books Ltd, Bodmin, Cornwall

Contents

About the authors

LEAD AUTHORS

Dr Bor-Shiuan Cheng (鄭伯壎) is Professor of Organisational Behaviour and Chairperson in the Department of Psychology, the National Taiwan University. He received his BA (1975), his MS (1977) and his PhD (1985) in psychology from the National Taiwan University. During his illustrious career he has received the National Science Council Outstanding Research Award and the Acer Best Scholar Award in Management Studies. He has also been postdoctoral fellow at the Institute of Industrial Relations, University of California at Berkeley, a visiting professor at the Judge Institute of Management Studies, University of Cambridge, UK and at the Euro-Asia and Comparative Management centre of INSEAD, France.

Professor Cheng's primary research interests include leadership, inter-organisational networks, and organisational culture in Chinese organisations. He is also interested in the topics of organisational development and change and employee and customer satisfaction. He is the author or co-author of many books and over fifty articles in learned journals, such as *Personnel Psychology*, the *Journal of International Business Studies*, *Organizational Science*, the *Journal of Social Behavior and Personality*, the *Journal of Cross-Cultural Psychology* and the *Chinese Journal of Psychology*. His most recent books are *Management in Taiwan and China*, 4 volumes (co-edited with Kuo-long Huang and Chien-Chin Kuo) (Yuan-Liou, 1998), *Organizational Culture: Analysis of Employee Level* (Yuan-Liou, 2001), *Organizational Behaviour Studies in Taiwan* (Laureate, 2003), *Leadership in Chinese Organizations* (Laureate, 2005) and *Paternalistic Leadership* (Hwatai, 2006). He is the current executive editor of *Indigenous Psychological Research in Chinese Societies*, and consulting editor of the *Asian Journal of Social Psychology* and *Management and Organization Review*. Professor Cheng is also a Director of the Philips Quality Education Foundation, and consults for major corporations in the fields of organisational culture, organisational development, leadership, employee satisfaction and customer satisfaction in both Taiwan and China. He is an acknowledged leader in his field, serving as director of Division of Industrial and Commercial Psychology, the Taiwanese Psychological Association, is a Life Member of Clare Hall of the University of Cambridge, UK and is on the advisory board of the Hang Lung Center for Organisational Research of the Hong Kong University of Science and Technology.

Dr Soo-Hung Terence Tsai (蔡舒恆) is Associate Director, MBA Programmes and a faculty member in Management (Business Strategy) at the Chinese University of Hong Kong. He received his BS in civil and environmental management from Cornell University, Ithaca, USA, his MS in environmental health management from Harvard University, Cambridge, MA, USA, and his PhD in Management Studies from St John's College, University of Cambridge, UK. Prior to joining CUHK, Professor Tsai was a Professor of Global Environment of Business at the Richard Ivey School of Business, the University of Western Ontario in Canada. His pre-Ivey career includes serving as a Rothmans Research Fellow in International Business at the Judge Business School and as a Research Fellow at Clare Hall, University of Cambridge, UK. He still holds the position of Senior Research Associate at Cambridge and is Visiting Professor of Management at Dalian University of Science and Technology in China.

Professor Tsai's research interests include multinational corporations, environmental management, organisational theory (environmental adaptive theories) and Chinese management. His most recent book *Corporate Environmentalism in China and Taiwan* was published by Palgrave-Macmillan in 2002. His other scholarly work has appeared in the *Journal of Management Studies*, *International Studies on Management and Organisation*, *Journal of General Management*, *Asian Case Research Journal*, *Business Strategy and the Environment*, the *Case Journal of Dalian University of Science and Technology*, *The Globe and Mail*, *Recruit*, the *Harvard China Review* and the *Sun Yat-sen Management Review*. Professor Tsai has also been a guest speaker at various international management conferences and serves as an advisor and a committee member of the United Nations Environmental Programme, the Academy of Management, the Greening of Industry Network and the Republic of China Strategic Alliance Society. Professor Tsai is currently on the editorial board of *Business and the Natural Environment* and the *Case Journal of Dalian University of Science and Technology*.

Professor Tsai is a consultant to the International Business Ethics Institute and was a full-time consultant at the General Electric Company, CH2M HILL Inc. Melzer Management Consulting (Germany) and has acted as an advisor to the National Environmental Protection Agency of China, the Environmental Protection Administration of Taiwan and the Massachusetts Department of Environmental Protection. His past consulting clients include organisations such as Acer, Dow Chemicals, Du Pont, Exxon, Hoest, ICI, Merck, Pepsi Cola, Philips and UMC.

Professor Tsai has lived and taught in Canada, China (including Hong Kong), Singapore, Taiwan, the UK and the USA.

CONTRIBUTING AUTHORS

Ms Lena Croft (簡曼萍) is a PhD student at the Chinese University of Hong Kong. Her research interest is foreign direct investment with special focus on the Chinese energy sector. Ms Croft is currently a subcommittee member (research and government liaison) of the Energy Committee, American Chamber of Commerce in Hong Kong.

Ms Donna Everatt holds an MBA from the Richard Ivey School of Business, the University of Western Ontario, Canada, and was employed as an Asian case writer at the same institution.

Dr Min-ping Huang (黃敏萍) is Associate Professor in the Department of Business Administration, Yuan Ze University, Taiwan. She received her PhD in business administration from the National Taiwan University, Taiwan. She is currently working on a series of papers examining how team composition, team structure and team leadership affect intra-team process and team effectiveness. She is also interested in how charismatic leadership affects employees' value fit and their effectiveness. Her most recent papers have been published in the *Asian Journal of Social Psychology*, *International Journal of Manpower*, *Journal of Psychology in Chinese Societies*, *Indigenous Psychological Research in Chinese Societies* and *Journal of Management*.

Dr Chin-kang Jen (任金剛) is Associate Professor of the Institute of Human Resource Management, the Yat-sen School of Management, the National Sun Yat-sen University. He received his PhD in organisational behaviour and human resource management from the National Taiwan University. His current research interests include organisational culture, trust, loyalty and organisations in the Chinese context. He has consulted for numerous public and private sector firms in Taiwan including Acer, China Steel, ITRI, Philips Taiwan, TSMC and UMC.

Dr Chia-wu Lin (林家五) is Associate Professor of Organisational Behaviour and Human Resource Management, at the Department of Business Management, National Dong Hwa University, Hualien, Taiwan. He has a PhD from the National Taiwan University, Taiwan. His research focuses on organisational identity, 'sense-making' in organisations and decision-making under uncertainty. His research articles have appeared in the *Journal of Management* (Taiwan), the *Bulletin of the Institute of Ethnology Academia Sinica* (Taiwan) and the *Journal of Human Resource Management* (Taiwan).

Mr Tsung-yu Wu (吳宗祐**)** is Assistant Professor of the Department of Psychology, Soochow University. He received his Bachelor Degree in international trade and his Masters degree and PhD in industrial psychology from the National Taiwan University. He is also a licensed clinical psychologist. His research interest is in stress and emotion at work. He is currently working on exploring how emotion at work affects employees' effectiveness and well-being. His most recent paper was published in *Indigenous Psychological Research in Chinese Society* and *Asian Journal of Social Psychology*.

Dr Chang-hui Zhou (周長輝**)** is Assistant Professor in Strategy and International Management at the Guanghua School of Management, Peking University, Beijing, China. He holds a BSc in Mechanical Engineering from Beijing University of Chemical Technology (1990), an MA in Economics from Renmin University, an MA in Economics from the University of Western Ontario, Canada and a PhD in Business Strategy from the Richard Ivey School of Business, the University of Western Ontario, Canada. His current research interests include innovation and knowledge management in multinational firms, mergers and acquisitions in China and the internationalisation of Chinese enterprises. Professor Zhou's papers have appeared in the *Asia Pacific Journal of Management* and the *International Studies of Management and Organization*.

TRANSLATORS

Ms Li-fang Chou (周麗芳**)** is Assistant Professor in the Department of Psychology, Kaohsiung Medical University, Kaohsiung, Taiwan. She received her Master's degree in Social Science in the Institute of Agricultural Extension, and her PhD in Psychology from the National Taiwan University. She is currently interested in social networks and *Guanxi* of intra-organisation. She is also interested in the issue of paternalistic leadership and teams. Her most recent papers have been published in the *Asian Journal of Social Psychology*, the *Journal of Psychology in Chinese Societies* and *Indigenous Psychological Research in Chinese Societies*.

Ms Hsiu-hua Sophia Hu (胡秀華) is Assistant Professor in Management at the Min Chuan University, Taipei. She received her PhD from the National Taiwan University, Taiwan. She has been Secretary General, Asia Pacific Association for Business Administration since 1998 and a Senior HR Consultant, Watson Wyatt, Taiwan Branch, from 1991 to 2000.

Ms Yueh-Ysen Lin (林玥岑) is Assistant Professor in the Department of Business Administration, Yuan Ze University, Taiwan. She holds an MA in HRIR (Master of Arts in Human Resource and Industrial Relations) and a PhD from the University of Minnesota-Twin Cities.

Mr Chun-Jung Tseng (曾春榮) is the Site Administrator of SITA, the world's leading provider of global information and telecommunication solutions to the air transport and related industries, in the CKS International airport, Taoyuan, Taiwan. Mr Tseng received his Masters degree in Social Science in the Institute of Agricultural Extension, the National Taiwan University. He is now a professional translator, with recent works on *Visual Basic 6 In Record Time*, the *MCSD: Visual Basic 6 Distributed Applications Study Guide*, the *CCDA CISCO Certified Design Associate Study Guide* and articles in computer science and psychology journals.

Preface

Whether we like it or not, we are already living in a globalised world. 'Globalisation' means that people are living in a world without distance – express transportation and instant communication are the norm. Globalisation also proves that each country's or region's development can reflect its own historical tradition and self-definition rather than simply following the developmental route of Western countries. For economic development, the capitalism/modernisation model is not a 'one size fits all' solution. While capitalism has varied modes, modernisation also has various paths to progress. Since the modernisation model is multi-faceted, globalisation has the role of stimulating communication between diverse 'modernisations' in order to create economic development.

The writing of this book is based on the thesis of such intercommunication, and attempts to contrast the success of the semiconductor industry in Taiwan with its development in other countries and regions. Although the semiconductor industry in Taiwan has since 1970 enjoyed rapid growth and some significant success in worldwide competition, there has still been little investigation into the key factors driving this success. The few existing discussions generally focus on macro issues of national politics and industrial environment rather than micro issues of organisational strategies. As a result, our knowledge on the organisational factors underpinning the semiconductor industry in Taiwan is still very limited.

As the Chinese idiom says, 'Chih Nan Hsing Yi' – To know is always easier than to take action. For most scholars what is lacking when doing research is not awareness of the problem but how practically to tackle it. At Professor Terence Tsai's invitation, Professor Paul Beamish from the Richard Ivey School of Business of the University of Western Ontario in Canada visited Taiwan in 1997, which motivated us to begin our investigation. After visiting some high-tech electronic companies such as Philips Taiwan and United Microelectronics Corporation (UMC), Professor Beamish agreed that Taiwan's success story deserved to be disseminated worldwide. He sent a special case study team to Taiwan to visit some corporations, and four cases regarding high-tech electronic companies in Taiwan were published; that on Acer received the Ivey Award as one of the top ten most popular cases in 2002 and 2004. We then wrote case studies on the Industrial Technology Research Institute (ITRI), the Macronix

International Corporation (MXIC), the Taiwan Semiconductor Manufacturing Corporation (TSMC), Applied Materials Taiwan (AMT), the United Test Centre (UTC) and Philips Semiconductors Kaohsiung (PSK), to show the practical application of vertical integration for job-sharing in high-tech industries in Taiwan, and to elucidate how the semiconductor industry in Taiwan had achieved its success. Professor Beamish's insight and action thus made a seminal contribution to the successful completion of this book.

This book took some five years to complete, and we want here to show our sincere appreciation to those people and departments that were supportive and encouraging in our research. First, we want to thank those supervisors and organisations that facilitated our visits and interviews, including Kung Wang and Jen-Tsung Pan (Hsinchu Science Park), Ta-Hsien Lo (ITRI), Y.S. Tan and Y.L. Lin (MXIC), David Wang and Arthur Tsao (TSMC), Andy Chang, E.N. Chou and Jeff Ma (UMC), I.K. Yin and Rund van der Linden (Philips Semiconductors), C.C. Tsai, J.L. Yang and Monica Chang (UTC), M.Y. Lin, George Hsu, B.Y. Chang and S.C. Lin (Acer). Second, we want to thank Professor Yau-De Wang, who introduced our team and the interviewees to each other. Third, we want to thank Professor Chong-Jen Chuang, Dr Yueh-Tsen Lin and all the assistants and graduate students of the Industrial and Commercial Psychology Research Unit at the National Taiwan University, who helped us with impressive interviewing and manuscript transcribing skills.

I wish finally to express gratitude from the bottom of my heart to all the co-authors, who spared no effort in either travelling overseas for interviews or working on generating ideas. Without their diligent and tireless work it would have been impossible to complete this book. The dedication of Li-Fang Chou and Yu-Jen Wang, from the National Taiwan University and Mary Tuen-Yi Au, from the Chinese University of Hong Kong, established a crucial paradigm for cross-national team cooperation. Having revealed the legendary success of the semiconductor industry in Taiwan, our hope is that the publication of this book will act as the first step to open a window of communication between multi-faceted modernisations.

Bor-Shiuan Cheng

Acknowledgements

This book is a culmination of our close to seven-year effort in tracing the development of high-tech industry in Taiwan. The idea was first conceived at the Judge Business School of the University of Cambridge where the authors first made encounters. The foundation of the book was laid at Ivey School of Business of the University of Western Ontario in Canada where Professor Paul Beamish introduced us to Edward Elgar Publishing.

Many people, too many to mention all of their names, have rendered their valuable help. Our co-authors have been most kind in being patient and relentlessly updating data and concepts. Our research assistants (Richard, Lena, Mary and Rhonda, among others) have devoted much effort in cross-validating data and 'coaching' us along. Our undergraduate, MBA, EMBA, PhD and masters students in Canada, Hong Kong and Taiwan have been constant sources of inspiration. We must confess to not being able to finish the manuscript without the support of our family members, friends and loved ones.

Last but not the least, we must also thank Hong Kong SAR Research Grant Council (CUHK 4311/01H) and Asia Pacific Institute of Business of the Chinese University of Hong Kong for providing us with grants to study this pollution intensive semiconductor industry. We are also particularly indebted to our institutions of affiliation, namely, Department of Management, Faculty of Business Administration of the Chinese University of Hong Kong and Department of Psychology, College of Science of the National University of Taiwan.

Terence Tsai, PhD
Bor-Shiuan Cheng, PhD

Abbreviations

ACRS	Annual Customer Relation Survey Table
AI	Artifical Intelligence
AMC	Applied Materials China
AME	Applied Materials Europe
AMJ	Applied Materials Japan
AMK	Applied Materials Korea
AMPI	Advanced Microelectronic Products Inc.
AMSEA	Applied Materials South East Asia
AMT	Applied Materials Taiwan
APOC	Asia-Pacific Operation Centre
ASE	Advanced Semiconductor Engineering Group
ASI	Asian Semiconductor Inc.
ASICS	Application Specific Integrated Circuit
ATM	Asynchronous Transfer Mode
BGA	Ball Grid Array
BiCMOS	Bi Complementary Metal Oxide Semiconductor
CEPA	Closer Economic Partnership Agreement
CIP	Continuous Improvement Project
CMOS	Complementary Metal Oxide Semiconductor
CMP	Chemical Machine Polishing
CRT	Cathode Ray Tube
CSP	Chip Scale Package
CVD	Chemical Vapour Deposition
CYCU	Chung Yuan Christian University
DRAM	Dynamic Random Access Memory
ELS	Executive Listening Session
EP	Extended Partition
EPROM	Erasable Programmable Read Only Memory
EPZ	Export Processing Zone
ERSO	Electronics Research and Service Orgainsation
FDI	Foreign Direct Investment
FPD	Flat Panel Display
HMC	Hualon Microelectronics Corp.
HRM	Human Resources Management
HSIP	Hsinchu Science-based Industrial Park

IBSS	Installed Base Support Survey
IDM	Integrated Device Manufacture
IDP	Individual Development Plan
IMEC	Inter-university MicroElectronics Center
IMP	International Material Research Company
IO	Input/Output
IP	Intellectual Property
IPG	Information Products Group
IPP	Intellectual Property Protection
ISP	Integrated Solution Provider
ISP	Integrated Support Package
ITRI	Industrial Technology Research Institute
ITRS	International Technology Roadmap for Semiconductors
JDP	Joint Develpoment Project
JIT	Just-in-time
JV	Joint Venture
LDC	Less Developed Country
MCP	Multi-Chip Packaging
MDEA	Ministry of Economic Affairs
MES	Manufacturing Execution System
MLCC	Multi-Layer Chip Capacitor
MOS	Metal Oxcide Semiconductor
MXIC	Macromix International Corporation
NASDAQ	National Association of Securities Dealers Automated Quotation
NCKU	National Cheng Kung University
NCTU	National Chaio-Tung University
NKK	NKK Corporation
NRCs	Non-return Engineering Charges
NTC	Naya Technology Corporation
NTHU	National Chaio-Hua University
NTU	National Taiwan University
NVM	Non-volatile Memory
OEMs	Original Equipment Manufacturers
OES	Opto-Electronics System Laboratories
PDC	Process Diagnostics and Control
ProMOS	ProMOS Technologies
PSK	Philips Semiconductor Kaohsiung
PVD	Physical Vapour Desposition
QC	Quality Circle
QFP	Quad Flat Package
QRE	Quality Reliability

RBU	Regional Business Unit
RCA	Name of the company, but also term commonly referring to a specific wet cleaning recipe introduced by RCA in early 1970
ROC	Republic of China
ROI	Return on Investment
ROM	Read Only Memory
RTP	Rapid Thermal Processing
SBU	Strategic Business Unit
SCP	Structure–Conduct–Performance
SEZ	Special Economic Zone
SFC	Securities and Futures Commission
SIA	Semiconductor Industry Association
SIS	Silicon Integrated Systems Corp.
SMAIS	Space Media for Amenity, Intelligence & Image Systems
SMD	Surface-Mount Device
SMEs	Small and Medium-sized Enterprises
SMIC	Semiconductor Manufacturing International Corporation
SoC	System-on-a-Chip
SOE	State-owned Enterprises
SPIC	Siliconware Precision Industries Co., Ltd,
SRAM	Static Random Access Memory
SSMC	Systems on Silicon Manufacturing Company
STA	Strategic Technology Alliances
TAC	Technology Advisory Council
TCE	Transaction Cost Economics
TMC	Taiwan Mask Company
TMT	Touch Micro-system Technology
TQM	Total Quality Management
TSIA	Taiwan Semiconductor Industry Association
TSMC	Taiwan Semiconductor Manufacturing Corporation
TSOP	Thin Small Outline Package
UMC	United Microelectronics Corporation
UTC	United Test Centre
VIA	VIA Technologies, Inc.
VIS	Vanguard International Semiconductor
VLSI	Very Large Scale Integration
VST	Virtual Silicon Technology
VT	Video Terminal
WIP	Work in Progress
WSMC	Worldwide Semiconductor Manufacturing Corp.
WTO	World Trade Organisation

1. 'Dragon appearing in the field': the legend of the semiconductor industry in Taiwan

Bor-Shiuan Cheng

PROUD ACHIEVEMENT: THE EMERGENCE OF THE INFORMATION INDUSTRY IN TAIWAN

To judge from its size, Taiwan, an island of about 30 000 km^2, is only a tiny dot on the globe. Its size, almost equal to that of the Netherlands, is about one seventh the size of the United Kingdom, one 250th the size of the United States, one 260th that of China, and cannot even be seen on a world map. Taiwan had nonetheless achieved a production value of US$21 000 million by 1999 and become the third biggest manufacturing centre of information hardware with over 40 computer-related products ranked first in the world. These products include scanners, monitors, motherboards, desktops and notebooks, and key components such as computer chips. Considering its size, Taiwan's achievements in the computer industry are remarkable.

Following the development of the global information industry, the trend of Taiwan's information industry, though levelling out slightly in the early 1990s, has been gradually upward. Capitalising on a firm foundation built on the production of computer systems, components, and peripherals, industry latecomers outperformed their predecessors, expanding into liquid crystal display (LCD) monitors and chip production, as well as marketing and services. Together, the information industry has developed a strong capability in design and system engineering, and can provide foreign semiconductor and computer manufacturers with comprehensive services, from design, manufacturing, and marketing, to after-sales support and services. This is shown in a comment by Bill Gates, founder of Microsoft, at the 2000 IT World Congress in Taiwan, to the effect that Taiwan is very proficient in mass production, and thus has changed the face of computing. In stark contrast to other countries, Taiwan's triumph has made it one of the most successful stories in Asia (Dedrick and Kraemer,

2000). For instance, the Canadian government attempted to advance its semiconductor industry with budgets several times the size of Taiwan's in the 1970s, and Brazil in the 1980s prohibited foreign computer imports to give its computer industry a boost. Both proved to be failures (Addison, 2000).

How could Taiwan have had such an outstanding performance in the information industry? How did this industry in Taiwan obtain and maintain its competitive advantage? What was the secret of success? What kinds of roles did the government and manufacturers play during the development process? What insights can newcomers gain from earlier achievements? Though it might be thought that success is simply the result of being in the right place at the right time, luck cannot be the only reason. Success stories always attract researcher attention, for they provide much food for thought. However studies on the success of Taiwan's information industry are surprisingly limited, and in research studies done on Asian countries, Taiwan received very little attention. For example, in Dedrick and Kraemer's *Asia's Computer Challenge: Threat or Opportunity for the United States and the World?* (2000), the authors adopted a cross-national perspective to compare and examine reasons for the rapid rise of the computer industry in Asian countries, but allocated only half a chapter, out of a total eight chapters, to the study of Taiwan. Researchers have not paid much attention to Taiwan's information industry, so we are not very clear about the reasons behind its success.

Of the limited literature that does discuss Taiwan's information industry, most examines the development process from a macro point of view and ignores many critical micro factors. Researchers have usually adopted ideas about government policies for business and the global production system to describe the development of Taiwan's information industry. These include ideas about the government's centralised intervention policies and tough implementation regime, and whether Taiwan's information industry can be rapidly merged into the global production system. On the former point, researchers hold different opinions, with the debate extending even to the advantages and disadvantages of a market and a planned economy. Market economy advocates argue that Taiwan's government maintains a low level of intervention in industrial development. That is, the Taiwan government, except for providing a stable macroeconomic environment and establishing basic infrastructure (that is education, research and development (R&D), telecommunications infrastructure, electric power and transportation systems), does not overuse industrial policy to interfere in market functioning, thus allowing more natural development. Planned economy advocates have a different view, believing that the stellar performance of Taiwan's information industry is the result of the government's selective intervention and strong assistance policies. Because of government intervention in reducing the risk of failure, local investors are more willing to pour capital into a high-risk sunrise information industry. Also, with an abundance of resources, companies are able to rapidly become more competitive and successful. Surprisingly, both market economists and planned economists consider the Taiwan experience as good evidence to support their own arguments. Thus, depending on which perspective one takes, entirely different conclusions can be reached.

On the other hand, advocates of a global production system mainly attribute the rise of Taiwan's information industry to the rapid integration with the global information production system, rather than the government's intervention in industry. The PC revolution of the 1980s changed the industrial structure from vertical integration (the mainframe era) to horizontal distribution (the era of the Internet economy). This change, in turn, formed the globalised production system and led to all kinds of manufacturing activities around the world. The close connection between Taiwan and the global production system under this industrial structure facilitated utilisation by Taiwan's information industry of national and local competitive advantage to improve its position in the global supply chain. One of the reasons contributing to Taiwan's success is thus that the government and manufacturers clearly understood the development trend of the global information industry. The other reason is that the government encouraged investment of foreign capital, endeavoured to transfer foreign technology to Taiwan, and assisted domestic enterprises in developing core competencies in the global division of labour. Manufacturers' rapid discovery and exploitation of opportunities is yet another important reason for Taiwan's success. However there are still countries that adopted different strategies but also succeeded in rapidly merging into the worldwide market and taking prominent roles in the global supply chain. For example, Japan protects the domestic market and Korea encourages large conglomerates (*chaebols*) to develop their industry. Both countries have adopted different strategies but are still leaders in the information industry (Hamilton, 1997; Dedrick and Kraemer, 2000).

Putting aside the disagreements and arguments between researchers, macro explanations seem reasonable in illustrating the high-level factors contributing to the success of Taiwan's information industry. Yet overemphasising this perspective can lead one to miss important micro factors. These factors include decades of formation of interorganisational networking production systems in Taiwan; an entrepreneurial spirit that values being the leader ('better to be the head of a chicken than the tail of a horse'); manufacturers' abilities to respond to market demand rapidly and accurately; and possession of a global network of interpersonal relationships.

In other words, besides macro explanations, seeing the evolution through the eyes of manufacturers and organisations can help uncover the reasons for the rapid rise of Taiwan's information industry more clearly. More specifically, an industry's success is mainly dependent upon manufacturers' and organisations' performance and coordination. The global environment and government policies, though important, simply give them an impetus. After all, organisations and manufacturers are the real participants in industry, and thus their operating strategy, organisational design and technological development are closely linked to industrial development. Research focusing on organisations and manufacturers is therefore important, given that it not only helps to uncover the reason for Taiwan's successful development of information industry in a more detailed way, but also makes up for the lack of high-level structural analysis (for example, nation theory) to form a more complete framework of the development mechanism of the information industry (Hamilton, 1997).

In addition to the exploration of macro factors, past research on Taiwan's information industry has tended to study Taiwan's computer industry, yet ignored the thriving semiconductor industry. Because the start of Taiwan's computer industry preceded other related industries, Taiwan is now the main manufacturing centre in the world, with a high market share for many products. In large part because of IBM's adoption of PC production strategies such as modularisation, open architecture and global resource searching, the structure of the global computer industry saw unprecedented change. By seizing a rare opportunity, Taiwan's computer makers succeeded in producing IBM-compatible computers and developed outstanding engineering abilities and highly flexible technological skills. This progress also facilitated the formation of an efficient link between computer manufacturing and supply, so that manufacturers could satisfy rapidly changing market demand at the lowest cost and fastest speed. Some large-scale computer makers, such as Acer (see Chapters 10 and 11), emerged at this time. It is thus no wonder that the computer industry became the research focus and guiding light of entrepreneurs.

Yet, was the development path of the semiconductor industry the same as the successful experience of the computer industry? Can conclusions drawn from the computer industry be utilised to explain the growth of the semiconductor industry? The answers to these questions are obviously 'no'. When comparing the two industries, the computer industry displays some special characteristics, while the semiconductor industry has had its own road to success. Many researchers agree that there are some very fundamental ways in which the growth of these two industries has differed. First, the government intervened in the two industries to a different extent. For the computer industry, the initial impetus was more or less a result of fortuitous history. The government's intervention was not really designed to spur industrial development but to destroy manufacturing of arcade games. As the government made it difficult for retailers to sell these games, thus effectively eliminating the market, the best way for manufacturers to survive was to shift production to the burgeoning area of PCs. In so doing, manufacturers accumulated a certain level of fabrication ability (Addison, 2000). Thus when IBM opened its manufacturing framework to public access, Taiwan naturally got its ticket to the PC market, and took an important position in the industrial chain. On the other hand, the development of Taiwan's semiconductor industry was another story. It was a result of the government's intentional cultivation and coordination of overseas Chinese and local entrepreneurs without much intervention from foreign investors (Hong, 1992; Mathew, 1995).

Second, the way of developing the required technology was different. Taiwan's computer manufacturing technology was introduced by foreign enterprises from America, Japan and Europe. In the 1960s those enterprises invested in consumer electronic products and components, but had expanded their scope to computer components by the 1970s. During that period a number of factories were established to play the roles of suppliers and subcontractors. Local producers endeavoured to learn manufacturing techniques and theory, and gradually transformed their role of subcontractor into producer of the entire computer. While the government did not provide much technical assistance, it proposed beneficial programmes and established processing zones to encourage investment in the electronics industry. In the 1980s most of the PCs were Taiwan-made but branded by multinational companies. By the 1990s, however, local businesses had a 70 per cent share in hardware production.

The introduction of semiconductor technology was completely different. In the 1970s foreign enterprises in Taiwan shifted only low-level techniques to Taiwan, such as packaging and testing, but kept high-level techniques solely in the hands of the parent companies. The government played an important role by establishing the Industrial Technology Research Institute (ITRI), in charge of the research and development of semiconductor technology as provided in the national budget. By 1995, the government was transferring successful technologies to the private sector. As the semiconductor industry expanded, foreign companies (including material supply, facility production, and integrated circuit (IC) design) started to invest heavily in the industry and laid down roots in Taiwan to cash in on the new opportunities (Liu, 1993; Wu and Shen, 1998).

If the computer industry was built up by imitation, the semiconductor industry used more innovation; if the computer industry followed the historic path, the semiconductor industry was built up step by step; if the computer industry was labour-intensive, the semiconductor industry was technique-intensive. In other words, the story of the computer industry cannot appropriately describe the development path of the semiconductor industry. The development of Taiwan's semiconductor industry has its own unique features. It should thus be studied and depicted in a systematic and detailed way. The development process can thus provide important insights to policy makers, academic research groups and practical entrepreneurs around the world.

THE DEVELOPMENT BACKGROUND OF TAIWAN'S SEMICONDUCTOR INDUSTRY

The transistor, invented by the Nobel Laureate in Physics, William Schockley, in the US Bell Laboratories in 1947, was undoubtedly one of the most important discoveries in the history of scientific technology in the twentieth century, an invention that also rapidly changed humankind's life. Later, Bob Noyce of Fairchild Semiconductors invented the surface-mount technology of silicon chips and made it possible to integrate many transistors onto a single chip. After 1959, the technique of integrated circuits began to progress at a tremendous rate, and the number of transistors put on a chip doubled every year. The decrease in cost led to the rapid development of a semiconductor industry (Turton, 1996). At the start of the new millennium, the semiconductor industry is still a shining star among technological industries.

In the early stages American chip makers virtually monopolised the semiconductor market, and the situation stayed unchanged until Japanese producers began to join in the 1970s. By the end of the 1970s the American makers' market share had decreased from 100 per cent to 65 per cent, while Japanese makers took 25 per cent. In the mid-1980s, the Japanese makers' market share increased to 46 per cent and they had taken over the new 256K Dynamic Random Access Memory (DRAM) market. The market share of American firms decreased to 43 per cent, leading to unprecedented financial losses for makers in the 'Silicon Village'. By the early 1990s, Japanese makers, as the market leaders, supplied 48 per cent of the global market and ranked first in the world: the American makers' market share had decreased to 36 per cent (Wu and Shen, 1998). In other words competition in the semiconductor industry has been gradually intensifying: American producers were no longer the powerful manipulators of the market they had once been.

To respond to the strong competitive challenge from the Japanese, American manufacturers adopted a strategy to set up partners worldwide by shifting production processes with low-level technology but high labour cost, such as packaging and testing, to some Asian countries. By 1970 American manufacturers had built 46 packaging companies around Asia: 11 in Malaysia, nine in Korea, nine in Singapore, eight in Hong Kong, three in Taiwan and six in other countries. Taking Taiwan as an example, American companies investing in Taiwan included General Instruments, RCA and Texas Instruments. Philips from Europe was yet another company to adopt the strategy. These multinational companies introduced techniques of IC packaging, testing and quality control to Taiwan. The initial development of Taiwan's semiconductor industry was thus similar to other Asian countries, serving as an overseas station for packaging and testing for American and European companies. However later development was very different.

Some researchers declared that the development of Taiwan's semiconductor industry went from packaging, positioned at the back-end of semiconductor production, to producing, positioned at the front-end. This is the opposite of the development from the front-end stage to the back-end stage in advanced countries (Mathew, 1995). The claim, however, is wrong. In fact, foreign companies' investment in the packaging process in Taiwan did not really help in developing upward to semiconductor production, and did not contribute to Taiwan's semiconductor development as much as is commonly thought. Taiwan's IC production was in fact mainly the result of the government's strong support in R&D. The technical development of packaging actually progressed at the same pace as manufacturing, rather than more quickly. When foreign investment focused on semiconductor packaging and testing, the government endeavoured to research the production and design of semiconductors, and then transferred the technology to the local producers. Owing to the vigorous entrepreneurship culture in Taiwan, the semiconductor industry rapidly gained a place in the global semiconductor supply chain and this, in turn, led to the wave of investment in wafer fabrication. When the technology of semiconductor production reached maturity, Taiwan's companies expanded after obtaining technical and financial support either upward to the design level or downward to packaging/testing. Compared with the government's contribution, foreign investment was a minor factor. Lu (2001) described the progress as follows.

> In the 1970s Taiwan did not receive much attention from foreign IC makers, not to mention willingness to invest or build branches on the island. This seeming disadvantage actually provided Taiwan with nearly 20 years to build a solid foundation in technology, operations and marketing. Founded by the government, ITRI was widely thought to be a main influence in helping the industry success during the time. It first spun off UMC (see Chapter 6) and proved that Taiwan had ability in IC mass production. Then the 'Very Large Scale Integration (VLSI) Technology Development' project helped establish the first

> 6-inch fabrication laboratory in Taiwan, and led to the establishment of TSMC (see Chapter 5). As the Sub-micron Project was proved to be a success by the end of 1994, Taiwan became the leader of mass production of 8-inch wafers and independently developed their own memory technology. From then, Taiwan took a prominent position on the stage of world semiconductor manufacturing. Europe, America and Japan started to value Taiwan's strength, and were willing to negotiate projects featuring significant technological transfer. Also joint ventures between Taiwanese and foreign companies surged at the same time.

Since Taiwan took only ten years to develop its semiconductor industry and to take a position in the global industrial chain, the successful rise of Taiwan's IC industry raises several important questions for researchers. How did Taiwan's semiconductor industry develop? What facets of the achievement can Taiwan take credit for? In 1989 there were already six local fabrication plants in Taiwan; however, the total production value was only US$22.4 billion, one-fourth the amount of Advanced Micro Devices in the United States. Even Klaus Wiemer, President of TSMC, when questioned about the prospects for Taiwan's wafer production, said in a speech given in September 1990: 'I doubt the infrastructure requirements of a world class IC industry are even understood by many [in Taiwan]' (Addison, 2000, p. 82). Yet the whole situation changed in the early 1990s. In 1990 there were only eight manufacturers in Taiwan, with the 0.8 microns process and sales revenues totalling NT$9.1 billion. By 2000 the number of manufacturers had increased to 16. The technical level had improved to the world-class 0.13 microns technology, and the annual sales revenues had increased to NT$468.6 billion (Index of Taiwan's semiconductor industry, see Table 1.1). With a total production value over NT$714.4 billion, Taiwan is now the fourth biggest country of semiconductor supply in the world, the level of annual sales revenue is 50 times higher than it was in 1990, and production ability has advanced several generations.

Hidden behind these outstanding achievements, however, are painstaking efforts and unusual providence, which are worth discussing case by case. Growing out of nothing, Taiwan's semiconductor industry has been developing for 25 years. The region of the Hsinchu Science-based Industrial Park (HSIP, see Chapter 3), once a neglected graveyard, has now become a world-class science park with gleaming new facilities. It is thus worthwhile to try and track down the guiding spirit of Taiwan's semiconductor industry in recent history.

This chapter separates the development of Taiwan's semiconductor industry into four distinct and critical periods. The first stage is the period of *foreign investment*, which started in 1966 when General Instruments built a chip packaging factory in Taiwan, and continued until 1974, when

Table 1.1 Statistics of Taiwan's semiconductor manufacturing industry, 1987–2000

	1987	1988	1989	1990	1991	1992	1993	1994	1995	1996	1997	1998	1999	2000
Number of firms	4*	6	6	8	10	10	10	11	12	15	17	20	21	16
Sales revenue (NT$ billion)	3.8*	4.4	7.6	9.08	16.79	23.46	41.5	70.0	119.3	125.6	153.2	169.4	264.9	468.6
Growth rate (%)	5.5	15.8	72.7	19.5	84.9	39.7	76.9	68.7	70.4	5.3	22.0	10.6	56.4	76.9
Production ability (μm)	2.0	1.5	1.2	1.0	0.8	0.8	0.6	0.5	0.45	0.35	0.3	0.25	0.18	0.13
Import:Export ratio	49:51	56:44	45:55	59:41	64:36	54:46	47:53	37:63	33:67	32:68	44:56	45:55	50:50	34:66
Investment/Sales revenue (%)	101.3	152.3	64.4	120.3	116.9	40.7	25.4	37.6	139.4	68.6	78.5	74.4	71.4	65.9
R&D/Sales revenue (%)	–	–	10.0	8.8	9.6	7.9	6.3	4.7	5.1	5.4	7.7	10.5	7.0	5.3

Note: * Including the operation of *ERSO*.

Source: *Annual Reports on the Semiconductor Industry*, 1997, 2001.

the government started to dictate the direction of the semiconductor industry. Investment from multinational companies in new factories and the transference of packaging techniques were two features of this period.

The second stage, from 1975 to 1984, was the period of *government R&D*. In 1975 ITRI's Institute of Electronics helped to jump-start the localisation of the industry with the introduction of RCA's semiconductor production process. This period ended with the founding of UMC and many other IC design companies. During this time government-led R&D in semiconductor production provided necessary financial support, and transferred techniques to local enterprises.

The third stage, from 1985 to 1994, was the period of *private enterprise expansion*. The period kicked off as UMC was listed on the stock exchange in 1985, and concluded with the success of submicron technology in 1994. At the time the development of Taiwan's IC technology was gradually shifting from government units to the private sector and growing exponentially.

The fourth stage, from 1995 to 2000, was the period of *investment upsurge*. In 1995 Taiwan's semiconductor industry succeeded in the mass production of 8-inch wafers and planned to invest in 24 fabrication factories. This period ended with the announcement in 2000 of the establishment of a number of 12-inch wafer fabrication companies. By that time conditions had become ripe for a deluge of private investment. Wafer foundries became the so-called 'star strategy' of the global semiconductor industry. Meanwhile, given the private sector's independence and self-sufficiency in R&D, the government began to reduce its involvement in the industry.

INDUSTRIAL DEVELOPMENT: HISTORY OF TAIWAN'S SEMICONDUCTOR INDUSTRY

Period I: Foreign Investment (1966–74)

The 1970s was a competitive era for America's top ten semiconductor companies. They started to explore niche markets in response to strong pressure from Japan, as well as domestic competition from small-scale emerging companies (Saxenian, 1994). After Fairchild, America's second largest chip maker, began to open up pioneering operations overseas, Texas Instruments (the largest), General Instruments (the fourth largest), and RCA (the sixth largest) in turn established their own packaging and testing factories in Taiwan (rankings were based on data from 1965, see Table 1.2 for the world's top ten enterprises in the semiconductor industry from 1955 to 1995). In addition, Europe's largest chip maker, Philips, also voted for Taiwan when it set up an IC packaging homebase in Kaohsiung. Given that

Table 1.2 The world's top 10 firms in the semiconductor industry, 1955–95

Rank	1955 Vacuum tube	1955 Transistor	1965 Semiconductor	1975 Integrated circuit	1985 VLSI	1995 Submicron
1	RCA	Hughs	Texas Instrument	Texas Instrument	Motorola	Intel
2	Sylvania	Transitron	Fairchild	Fairchild	Texas Instrument	NEC
3	GE	Philco	Motorola	National	NEC	Toshiba
4	Raytheon	Sylvania	General Instrument	Intel	Hitachi	Hitachi
5	Westinghouse	Texas Instrument	GE	Motorola	National	Motorola
6	Amperex	GE	RCA	Rockwell	Toshiba	Samsung
7	National Video	RCA	Sprague	General Instrument	Intel	Texas Instrument
8	Rawland	Westinghouse	Philco	RCA	Philips	Fujitsu
9	Eimac	Motorola	Transitron	Philips	Fujitsu	Mitsubishi
10	Lansdale	Clevite	Raytheon	AMD	Fairchild	Philips

Note: GE – General Electric.

Source: Tushman and O'Reilly (1997).

labour costs in IC packaging are relatively high and transportation costs low, switching packaging to Asian countries where labour is cheap can lower costs and improve competitiveness. Compared with other Far Eastern areas in the 1970s, Taiwan had relatively cheap labour, flexible investment procedures, allowed duty-free raw material and equipment imports and had a good relationship with the United States. It thus attracted a stream of American investments.

Though foreign enterprises introduced semiconductor packaging technology to Taiwan and indirectly facilitated expansion of the industry, they did not help Taiwan form a vertically integrated semiconductor industry on the island. One of the reasons was that the semiconductor producers transferred only low-level techniques to overseas areas, keeping more advanced and value-added production technology at headquarters. Moreover once the cost advantage for having overseas stations vanished, the foreign companies had to find a way to avoid losing competitiveness. To some extent these overseas investments represented an opportunistic attitude; on the other hand, Taiwan's local companies were mostly small-scale, and could engage only in manufacturing packaging cases as original equipment manufacturers (OEMs), not to mention the innovation and creativity required to develop high technology. Since Taiwan's manufacturers were unable to develop packaging techniques independently, it was almost impossible to produce semiconductors on its own. Even though some investors (such as Fine Products Microelectronics Corporation) were exceedingly courageous in trying to do so, they ended in failure after only a brief trial period.

To put it another way, foreign companies' lack of willingness and local enterprises' lack of ability necessitated government intervention in steering a course for the new industry. In the light of this, Taiwan's government officials and overseas Chinese consultants (most were members of the Electronic Technical Advisory Committee), after extensive discussion, made an important decision. In 1973 the ITRI set up a new division, the Centre of Electronics Industry Development and Research (later changed to the Electronics Research and Service Organisation (ERSO)) to be in charge of the promotion of IC industry development and the introduction of RCA's IC design and manufacturing technology (Su, 1994; Shen, 1997). Every innovation has a problematic beginning, and the issue of whether Taiwan should promote and develop its own IC industry was the subject of vigorous debate. The focal points of the debate included questions such as: Did Taiwan really need an IC industry? Did Taiwan have sufficient resources and qualifications to develop its IC industry? Did Taiwan have enough international competitiveness to challenge other countries? Showing great wisdom, vision and decisiveness, key players put forward the vision that Taiwan could be more than a long-term manufacturing centre for low-cost products and, thus must upgrade its technology and reduce its dependence on multinational companies. 'By the late 1970s and early 1980s we had reached full employment and labour-intensive industries could no longer survive', said K.T. Li, the 'godfather' of Taiwan's science and technology (personal recollection). Taiwan chose semiconductor production as a strategic industry, and invested a huge amount of money in supporting its development (Addison, 2000).

Period II: Government Research and Development (1975–84)

After the government announced that IC manufacturing was to be a strategic industry and established a research institute to be responsible for it, the semiconductor industry began to develop lasting foundations. ERSO at ITRI carried out three important projects to solidify the industry's foundation. The first project was to set up a flagship IC factory by transferring from RCA the production process of 7-micron Metal Oxide Semiconductor (MOS) in 1975. The project, which came to fruition in 1977, successfully established several significant companies whose yield rate and performance were even better than RCA's American factories. In 1980 Taiwan's first homegrown IC manufacturer, United Microelectronics Corporation (UMC), was established with ITRI's assistance. Ironically while R&D was led by the government, UMC itself did not receive much support from private entrepreneurs and initially had difficulty in raising funds. Obviously most private entrepreneurs in Taiwan continued to hold a conservative view concerning the IC industry. As a result of a lack of external support, UMC drew resources from its employees, eventually becoming the company with the largest percentage of employee shareholding and with a so-called 'partnership culture' that sees all employees as contributors (Chou *et al.*, 2000). As its business flourished UMC became the world's second largest wafer foundry. The project also cultivated many important technical talents and entrepreneurs to provide the necessary human capital for the establishment of new enterprises.

The purpose of the second project was to introduce mask-repeating technology to improve the efficiency of the R&D of IC products by curtailing between one and two weeks of production time. This success also facilitated the establishment of the Taiwan Mask Company (TMC) in 1981 and implementation of the third project, the VLSI project. The goal of this third project was to develop automation ability of production technology and design of VLSI. Before reaching the conclusion of the project, there were arguments about whether Taiwan should focus on the development of design-intensive products (such as Application Specific Integrated Circuit – ASICS) or standardised products (such as DRAM). In the end, the decision was made to continue developing design-intensive products as they best fitted the scale, technology and capacities of Taiwan's enterprises. This decision has largely been proved correct, given that it exploits Taiwan's competitive advantages in the production of diverse, low-volume products and the industrial strategy towards vertical disintegration. This production strategy is very different from Korea's, which puts an emphasis on mastering mass production of standardised products (Dedrick and Kraemer, 2000). In addition this project brought opportunities for cooperation between overseas Chinese and Taiwan's institutes. Many America-based small-scale companies started to expand their business to Taiwan. After the establishment of HSIP, Taiwan became a strategic partner of 'Silicon Valley' in the United States, leading to frequent interactions between the Bay area and Taiwan (Saxenian, 1997). In fact, HSIP imitated the Stanford Research Park and drew heavily on the successes of the 'Silicon Valley'.

When the VLSI project proved to be a success, the government used the research to spin off a third chip maker, Taiwan Semiconductor Manufacturing Corp. (TSMC, founded in 1986, see Chapter 5). In fact between 1982 and 1985 other knowledge-intensive but low-cost IC design companies were spun off from ERSO. In other words, after the government set up the research institute, numerous technologies were researched, developed and transferred to the local sector over a period of ten years. Of the technologies, IC manufacturing techniques and IC design rules had the biggest impact on the island's economy. It is worthwhile noting that the government did not intervene in the mass production of the products. They kept a focus on R&D, but empowered the local private sector to be fully in charge of the production and commercialisation process. The fact that the government concentrated on research in promoting Taiwan, and transferred the technology to the private sector instead of establishing state-operated businesses, is an important part of the development of Taiwan's semiconductor industry.

Period III: Rise of Private Enterprise (1985–94)

As government R&D brought new technologies to maturity and transferred them to the private sector, profits at companies like UMC – ranked first of Taiwan's top 500 corporations in 1983 and staying near the top ever since – caused more and more local enterprises to gain confidence in the semiconductor industry and pour capital into the burgeoning sector. Following the establishment of TSMC, enterprises entering the field of semiconductor production in turn included Da Wang Electrics (1986), HMC (Hualon Microelectronics Corp.), AMPI (Advanced Microelectronic Products Inc.), Winbond Electronics Corporation (all 1987), Holtek Semiconductor Inc. (1988), MXIC (Macronix International Co. Ltd, 1989), Texas Instruments-Acer Inc. (1990), Mosel Vitelic Inc. (1991). The number of manufacturers climaxed at 21 in 1999, dropping to 16 by 2000 after consolidation. After the establishment of Syntekt Semiconductors (1982), He Teh Integrated Circuits (1983), Chip Design Technology, and Proton (1985), the number of IC design companies rapidly increased to 65 in 1994, with annual sales of NT$12.4 billion. The number is still increasing (there were 140 companies by 2000, with annual sales revenues of NT$115.2 billion, see Table 1.3). Meanwhile, the flourishing semiconductor industry also drove the growth of the packaging industry. By 1994 there were 18 package enterprises totalling annual sales revenues of NT$19.5 billion (NT$97.8 billion in 2000, see the *Annual Report of the Semiconductor Industry, 2001*).

Of the rising stars, some companies were spun off from ERSO at ITRI, others were built with the combined capital of American, Chinese and Taiwanese investment, others still were established through cooperation between Taiwan companies and multinational enterprises (Wu and Shen, 1998). After 1990 coordination among local businesses was common. The corporations included those vertically disintegrated in technology, marketing and manufacturing. In terms of technical cooperation, there were two main vehicles: either through technological cooperation with other countries, or cooperation between local manufacturers and foreign enterprises to launch new technologies. The purpose of market cooperation was to expand market share, using strategies such as mutual investment, united R&D, joint ventures and participation in industrial-standard alliances. In addition, the vertical disintegration within the industry was formed by adopting strategic alliances to integrate low- and high-margin processes and reshaping the interorganisational network. According to the analysis of Lu *et al.* (1997), Taiwan's IC manufacturers were grouped into nine main networking divisions by 1996. Each network includes about 8–15 companies with different functions, such as IC making, IC design, IC packaging, IC testing and IC marketing.

During this period, ITRI also started promoting the development of microelectronic technology (1988) and the submicron project (1990), both of which had succeeded by 1994 and spawned a fourth major chip maker, Vanguard International Semiconductors. It also drove an investment wave in 8-inch foundries, and forced Taiwan's semiconductor industry to enter a whole new era. From then on, Taiwan not only changed its role from a follower of its competitors, but also gradually became a world-class industry with an important and unchallenged position in the global supply chain. Taiwan thus freed itself from an image of imitator and copier, and

Table 1.3 Key statistics of Taiwan's IC design industry, 1987–2000

	1987	1988	1989	1990	1991	1992	1993	1994	1995	1996	1997	1998	1999	2000
Number of companies*	30	50	55	55	57	59	64	65	66	72	81	115	127	140
Sales revenues (NT$10 billion)	0.8	2.2	5.4	5.9	7.3	8.6	11.7	12.4	19.3	21.8	36.3	46.9	74.2	115.2
Growth rate (%)	42.8	175	145	9.3	24	30	36	6	56	13	67	29	58	55
Design Ability (Gate Number K)	–	10	20	30	30	30	60	60	65	400	–	–	–	–
Import/Export	–	26:74	57:43	64:36	49:51	50:50	46:54	65:35	61:39	64:36	52:48	57:43	62:38	59:41
Investment/Sales Revenue (%)	–	–	–	–	71.0	10.0	23.5	15.5	15.9	15.5	17.3	13.5	15.4	15.3
R&D/Sales Revenue (%)	–	–	–	9.4	9.9	10.1	9.5	10.0	12.2	9.5	8.8	9.4	8.9	9.3
R&D/Total Human Capital (%)	–	–	28	30	32.0	25.0	51.0	51.0	49.0	51.8	–	–	–	–
R&D average tenure (years)	–	–	–	4.0	4.5	4.7	5.4	5.9	5.0	5.3	–	–	–	–

Note: * Companies include professional IC design houses, foreign design centres, and design departments of system and IC manufacturing.

Source: *Annual Report of Semiconductor Industry*, 1997, 2001.

advanced to the position of innovative leader of technology for production processes and strategic management.

Period IV: The Upsurge of Investment (1995–2000)

Jerry Sanders, chief executive of Advanced Micro Devices, said: 'real men have fabs', given that 'owning silicon fabs is like having sharks in your swimming pool: it not only costs a surprising amount to feed [them] and cover expenses, but also consumes time and energy. Later, it ends up eating the owners' (Addison, 2000). Since fewer and fewer 'real men' existed in the new generation of managers in 'Silicon Valley', 'real men' emerged in Asia: Taiwan stopped feeding eels and advanced bravely without hesitation to foster sharks in the backyard pool. As a result, plans for 24 8-inch fabrication plants were announced in 1995. All were completed in rapid succession by 2000. The fact that the plants were built showed Taiwanese entrepreneurs' unstinting determination; these investments hastened the growth of Taiwan's semiconductor industry, and helped Taiwan to become one of the world's leading chip producers. Sales revenue in 1995 was only NT$119.3 billion, but had increased to NT$468.6 billion by 2000, a growth of almost 400 per cent. Having a shark in the backyard indeed changed the layout of the global semiconductor industry, and made the features of vertical disintegration more clear cut. Every field and production step within the IC industry is extremely specialised, and each manufacturer has a core competence which may be totally different from others. It is thus important to cooperate with manufacturers with the best competitive advantage to reduce the cost and exert the maximum value of the industrial chain. As roles in the global supply chain developed, many Taiwanese manufacturers became strategic partners of America's fabless design companies, given the strategy of focusing on wafer foundries adopted by Taiwan's enterprises. Since Taiwan's chip makers stressed low cost and high quality, and tended not to establish their own brands, many integrated-components enterprises were attracted to Taiwan, and tended to outsource and shift their wafer fabrication to Taiwan. In other words, wafer foundries became the main stream of Taiwan's semiconductor industry after 1995, first through being the OEMs of America's fabless chip makers, then by obtaining the patronage of America's large-scale integrated-component companies. Lastly, to lower investment costs Japanese semiconductor enterprises started to outsource to Taiwan's chip makers in 1997. In electric appliances and computers, Japan has always been faster than the United States to invest in Taiwan, and plays the role of precursor in spreading techniques. However this development model was not followed in the case of the semiconductor industry.

Since Europe, America and Japan in turn reduced investment in fabrication plants, plus the fact that more and more fabless chip makers tended to outsource their production to OEMs, Taiwan's status in global IC production was becoming solid. Taiwan not only attracted upstream providers of raw materials and equipment, but also brought in advanced investment, including Applied Materials (see Chapter 7) which set up a branch in Taiwan and soon became a critical link in the global supply chain. Each of the fabless producers and integrated-component companies, from the giant Intel to the smallest newcomer, had to partially outsource production to Taiwan, and no one could escape from it. Trouble at any Taiwanese manufacturer in the supply chain therefore caused a 'domino effect' and disturbed the global electronics industry. One of the most significant cases occurred after the Taiwanese earthquake of 21 September 1999, which led to a slide in stock prices of some of America's giants such as Dell and HP (Addison, 2000).

Besides the semiconductor industry's role as one of the important links in the world division of labour, Taiwan's IC design and lower-end packaging/testing also underwent rapid development. These fields not only formed some of the essential industries in Taiwan, but also fitted well with Taiwan's core competence in IC production. Taiwan's semiconductor industry, from design, production to packaging/testing, developed all round. Manufacturers possessed excellent abilities in independent research and design, and did not need to rely on government assistance. Overall production value totalled NT$714.4 billion in 2000, with NT$115.2 billion in IC design, NT$468.6 billion in IC production, NT$97.8 billion in IC packaging and NT$32.8 billion in IC testing (see Table 1.4). With some exceptions most of the sales were achieved by local companies, not foreign enterprises. In HSIP, for example, more than 80 per cent of the companies in this important production centre are local manufacturers, while foreign companies account for no more than 20 per cent.

Because of advances in private enterprise, now with full capacities in R&D, local companies began to raise questions when ITRI launched its fifth project, the deep submicron IC project. Local companies thought that development of IC production processes directed by the government should be concluded, given that many manufacturers were by then endeavouring to develop technology with better effect. Was it still necessary for the ITRI to continue to waste national funds on IC-related projects? No offshoot companies would be established even if the deep submicron project proved successful, or they would aggravate an already hyperactive competitive situation. Times had changed, and ITRI had fulfilled its responsibility in this phase. Should planning assistance for another fledgling industry be its next mission?

Table 1.4 Index of Taiwan's semiconductor industry, 1996–2000; growth 1999–2000

	1996	1997	1998	1999	2000	1999–2000
Industry production value (NT$ billion)	188.2	247.9	283.4	423.5	714.4	68.7%
IC design	21.8	36.3	46.9	74.2	115.2	55.3%
IC manufacturing	125.6	153.2	169.4	264.9	468.6	76.9%
Foundry value	56.0	84.2	93.8	140.4	296.6	111.3%
IC packaging	35.8	47.8	54.0	65.9	97.8	48.4%
Local packaging industry	25.2	36.2	42.0	54.9	83.8	52.6%
IC testing	5.0	10.6	13.1	18.5	32.8	77.3%
Product value (NT$ billion)	91.4	105.3	122.5	198.7	287.2	44.5%
Domestic sales (%)	3.9	4.7	4.97	5.47	5.39	–
Market value (NT$ billion)	203.6	235.5	274.4	345.7	506.5	46.5%

Source: *Annual Report of Semiconductor Industry, 2001.*

By 2000, Taiwan had become the fourth largest country for semiconductor production, and was producing about 12 per cent of the global chip market. Taiwan's chip makers, however, were not satisfied with this achievement and moved to build fabrication plants to use the newly developed 12-inch wafer production technology. Of the 35 12-inch wafer-making facilities that have been announced around the world, 19 will be locally owned and operated in Taiwan, and two will be owned by UMC's joint venture, but operated abroad. If these investments are carried out, then Taiwan will control two-thirds of the 12-inch wafer facilities around the world. Taiwan's share of the global semiconductor industry will reach 25 per cent. In response to the semiconductor manufacturers' expansion, the Taiwan government announced the establishment of the Tainan Science-based Industrial Park in 1997 in the hope of recreating HSIP's 'cluster effect' (Porter, 1998) and, in turn, improving Taiwan's semiconductor industry. Whether this strategy succeeds or not, the development of Taiwan's semiconductor industry from the 1970s to the turn of the millennium has been uniquely successful. This not only disproves the myth that less-advanced countries cannot develop a successful high-tech industry, but also disproves the thesis that developing countries necessarily need to rely on developed countries to develop their economies. From an industrial development perspective, the meteoric rise of Taiwan's semiconductor industry is miraculous. From an academic perspective, Taiwan's story is an important exception to all the textbook theories.

UNDER THE ICEBERG: INSTITUTES AND BUSINESSES IN TAIWAN'S SEMICONDUCTOR INDUSTRY

As stated by Porter (1990), research and analysis of national competitive advantage should focus on industry, rather than the nation. Similarly, the research and analysis of industrial competitive advantage should focus on firms and enterprises within the industrial value chain, rather than on industry itself. According to the theory of industrial value, industrial competitive advantage can be understood through the concept of *value chains* – a value chain provides a systemic structure to distinguish between firms' private activities, to explore the extent, types and clusters of these activities, and to examine their influence on competitive advantage (Porter, 1985; Gilbert and Strebel, 1991). More specifically, each firm in the value chain possesses its own competitive strategies and advantages. Thus through analysing each firm's competitive advantage and strategy, we can gain a better understanding of the reasons for industrial success (Evans and Wurster, 1997).

After more than 20 years of development, Taiwan's semiconductor industry has formed a highly competitive value chain, the competitive advantages being generated from organised activities among enterprises (Foss, 1996). Generally speaking, the value chain of the semiconductor industry can be separated into product design, mask production, wafer fabrication, packaging/testing, marketing and applications. Certainly it may not be precise enough to distinguish the whole production process in this way. The semiconductor industry is a highly integrated industry and depends on many disciplines such as electronics, electric machinery, chemistry, chemical engineering and information. Together the whole value chain is composed of product positioning, circuit design, layout design, simulation, mask design, lithography, etching, oxide/diffusion, implanting, deposition, wafer testing, dieing and packaging/testing. The finished goods are then transferred into the field through marketing and sales. Each step of the process is very delicate and closely connected to its upper and lower steps. Though these detailed processes can be roughly separated into design, wafer production and packaging/testing, it is more appropriate to separate the full process into design, masking, wafer production, packaging/testing, marketing and application.

Many businesses and enterprises are involved in each link of the value chain. For example, in 2000 there were 160 IC design companies, which could be grouped into two categories: fabless IC design companies such as Acer, VIA Technologies Inc. and Silicon Integrated Systems Corp. (SIS), and integrated device manufacturers with their own fabs, such as MXIC.

Also there were 16 wafer makers, including foundries such as TSMC and UMC, and makers with their own brands such as MXIC, Mosel Vitelic and Winbond. Other manufacturers in the value chain are shown in Figure 1.1.

Undoubtedly the industrial value chain is restricted by upper structural factors, and is dependent on support systems to provide a variety of resources (Hung and Whittington, 1997). In Taiwan's semiconductor industry, for example, government policy is obviously an important structural factor, as it provides not only technical support to drive the value chain's functioning, but also necessary resources (land, water and electricity, public facilities, for example) to support activity in the value chain. In Taiwan's semiconductor industry, institutes that execute the government's policy include ITRI, the Institute for Information Industry and the two Science-based Industrial Parks in Hsinchu and Tainan. The software, raw materials, facilities and peripherals that support the value chain's activities should also not be ignored, since these enterprises play a supporting role in providing necessary logistic support and smooth the functioning of the value chain. In other words, the semiconductor value chain, the supporting system and the government compose the complete system of Taiwan's semiconductor industry and, in turn, play an important role in the value chain of the worldwide information industry as a chip provider to satisfy the demands of the global market.

Basically the success story of Taiwan's semiconductor industry was written by many manufacturers and entrepreneurs working in a highly competitive and youthful market, led first by government involvement in research and design, provision of good 'infrastructure' and the introduction of technology from more sophisticated countries, and promotion of the industrial value chain. The government's policy, manufacturers' abilities (including the value chain and supporting system) and multinational technologies all possess their own unique place in the explanation of the industry's success. All kinds of macro structural factors as well as micro factors, such as manufacturers and people, complement each other, adding fuel to the development flames. More specifically, the government and the global industrial system initially shaped the embryonic structure of Taiwan's semiconductor industry; however it was the redesign of industry by the private sector that made Taiwan an important contributor to the global industry. The reasons for success are thus deeply embedded within the interaction between the manufacturers in the value chain, firms in complementary industries and the government, and analysts should carefully observe the interactions between the three groups within the development environment, since any explanation based on a single factor is probably insufficient. Questions to be asked include: How did the rapid change in markets and technology affect the government's and manufacturers' behaviour? As a

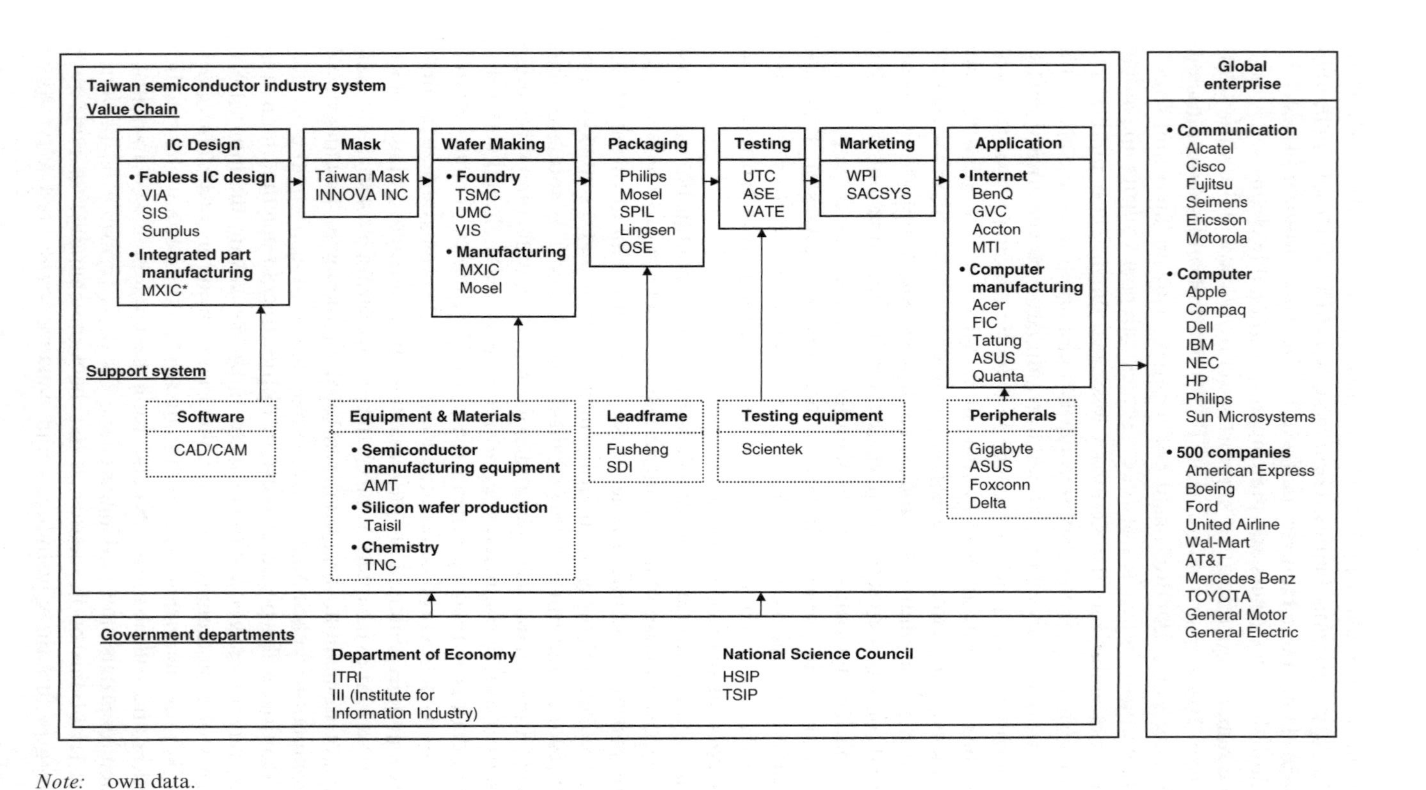

Note: own data.

Figure 1.1 *Taiwan's semiconductor industry and the global market*

policy maker, how did the government drive industrial development? How did the main enterprises adjust their management to maintain their solid status in the market? How did back-up businesses coordinate in order to maximise the effectiveness and performance of the value chain?

To summarise, we think that the reasons for the success of Taiwan's semiconductor industry are independent of government and institutions, the value chain's manufacturers and the companies in auxiliary industries. Through case analysis, the reasons for success can be revealed, and readers can identify with them. This book is developed on the basis of this perspective. In this chapter we summarised the development track and achievements of Taiwan's semiconductor industry, and proposed a theoretical value chain structure to help analyse the industry's success. Then according to the concept of the 'smile curve' (see Figure 10.2), we separate the book into three parts – first, a description of the 'smile curve': the development and fostering of the industry; then the front end of the 'smile curve': IC design and manufacturing; and finally the rear end of the 'smile curve': packaging/testing and application of IC.

Part I of this book examines the Industrial Technology Research Institute (ITRI) (Chapter 2) and the science parks in Taiwan: HSIP and TSIP (Chapter 3), to study the critical role the government played in the development of Taiwan's semiconductor industry. ITRI, similar to Fairchild, the pioneer of the American semiconductor industry, developed many important technologies and cultivated a 'talent pool' to help Taiwan rapidly expand its semiconductor industry. The HSIP (Hsinchu Science-based Industrial Park) geographically aggregated the main manufacturers and formed a platform for the so-called 'cluster effect'. It also became a portal to Taiwan's semiconductor industry for the external market.

Part II of this book includes three cases about the IC design and wafer manufacturing. One of the three cases, MXIC (Chapter 4) is a company with both IC design and wafer production capability, and its management strategy is quite different from Taiwan's core competence in foundries. TSMC and UMC (Chapters 5 and 6), the top two global foundries, are both the foci of discussion, since they have different strategies of management and compete with each other.

In addition, wafer production can be likened to gold digging. For a gold digger, relying on good digging tools is the only way to improve the efficiency of gold digging. Likewise, facilities needed for wafer fabrication is the key to the yield rate. We are therefore interested in the facility providers and will also consider the leading company of wafer facilities, Applied Materials Taiwan (Chapter 7).

Part III of this book focuses on the 'back-end' production process of Taiwan's IC industry, including IC packaging, testing and application.

In IC packaging Philips Taiwan (Chapter 8) is the company with the oldest brand and grew with Taiwan's IC industry. Philips also contributed to Taiwan's IC industry by introducing and shifting packaging technology to the island and has achieved solid performance. Part III also presents the case of the United Test Centre (UTC, Chapter 9), an outstanding latecomer in this field. The book also devotes some time to studying UTC's survival as a competitor to two other testing group groups, Advanced Semiconductor Engineering Group (ASE) and Siliconware Precision Industries Co. (SPIC). In addition, Acer, the most famous brand-building company in Taiwan's information industry, is studied as a case in IC application (Chapters 10 and 11). It is positioned at the end of the 'smile curve', and is distinguished by its globalisation strategy – moving from the countryside to the city. In other words, they first built a customer base in less developed markets, expanding only later into more developed areas.

Chapter 12 summarises the insights gained from all the cases in the book, and details the implications for industrial newcomers as well as global competitors by highlighting the characteristics and reasons for success of Taiwan's semiconductor industry. The authors also attempt to analyse the future trend of production of the global semiconductor industry and the implications this may have for both Taiwan's semiconductor industry and individual manufacturers. By understanding these trends industry watchers can continue to observe Taiwan's semiconductor industry and better predict its future success.

REFERENCES

Addison, C. (2000), *Silicon Shield: How Taiwan's High-Tech Industry will Protect the Island from a Military Attack by China.*

Chou, C.H., Tsai, T. and Cheng, B.S. (2000), 'Developing organizational capability from partnership and internal networking: the case of United Microelectronics Company (UMC) in Taiwan', *Asian Academy of Management Conference*, Singapore.

Dedrick, S. and Kraemer, K.L. (2000), *Asia's Computer Challenge: Threat or Opportunity for the United States and the World?*, New York: Oxford University Press.

Evans, P.B. and Wurster, T.S. (1997), 'Strategy and the new economics of information', *Harvard Business Review*, September–October, 71–82.

Foss, N.J. (1996), 'Research in strategy, economics, and Michael Porter', *Journal of Management Studies*, **33**(1), 1–24.

Gilbert, X. and Strebel, P. (1991), 'Developing competitive advantage', in H. Minzberg and J.B. Quinn (eds), *The Strategy Process: Concepts, Contexts, Cases*, Englewood Cliffs, NJ: Prentice-Hall, pp. 70–79.

Hamilton, G. (1997), 'Organization and market processes in Taiwan's capitalist economy', in M. Orru, G. Hamilton and N.W. Biggart (eds), *The Economic Organization of East Asian Capitalism*, Thousand Oaks, CA: Sage, pp. 237–93.

Hong, S.G. (1992), 'The Politics of Industrial Leapfrogging: The Semiconductor Industry in Taiwan and South Korea', PhD dissertation, Northwestern University.
Hung, S.C. and Whittington, R. (1997), 'Strategies and institutions: a pluralistic account of strategies in the Taiwanese computer industry', *Organization Studies*, **18**(4), 551–75.
Liu, C.Y. (1993), 'Government's role in developing high-tech industry: the case of Taiwan's semiconductor industry', *Technovation*, **13**, 299–309.
Lu, C.Y. (2001), *The Witness of Taiwan IC Industries*, Taipei: Sunbright.
Lu, Y.Y., Mao, C.J. and Cheng, Y.H. (1997), *Use of Resource Base Perspective in Exploring Competitive Strategy: The IC Manufacturing Industry*, Taipei: Hai-Tai.
Mathew, J. (1995), *High-Technology Industrialization in East Asia: The Case of the Semiconductor Industry in Taiwan and Korea*, Taipei: Chung-Hua Institute for Economic Research.
Porter, M. (1985), *Competitive Advantage: Creating and Sustaining Superior Performance*, New York: Free Press.
Porter, M. (1990), *Competitive Advantage of Nations*, London: Macmillan.
Porter, M. (1998), 'Clusters and the new economic of competition', *Harvard Business Review*, November–December.
Saxenian, A.L. (1994), *Regional Advantage*, Cambridge, MA: Harvard University Press.
Saxenian, A.L. (1997), 'Transnational entrepreneurs and regional industrialization: the Silicon valley–Hsinchu connection', *Conference Proceedings of Social Structure and Social Change*, Taipei: Academic Scinica.
Shen, R.C. (1997), 'The formative path of IC related industries in Taiwan: 1974–1982', *The Journal of Taiwan Bank Quarterly*, **48**(3), 55–82.
Su, L.Y. (1994), *The 20 Years of Elctronics Research Laboratory*, Taiwan, Hsin Chu: Taiwan Industrial Technology Research Institute.
Turton, R. (1996), *The Quantum Dot: A Journey into the Future of Microelectronics*, New York: Oxford University Press.
Wang, Z.F. (1999), *The Map of Electronic Industry in Taiwan*, Taiwan: Wealth Magazine.
Wu, S.W. and Shen, R.C. (1998), 'The formation and development of IC related industries in Taiwan', *Journal of Taiwan Industry Study*, **1**, 57–150.

2. 'The cradle of technology': the Industrial Technology Research Institute

Min-ping Huang

In front of the main building of the Industrial Technology Research Institute (ITRI) is the statue of a mother holding a child, symbolising ITRI's long-standing spirit and mission to promote and develop Taiwan's hi-tech industry. Since 1980, through the setting up of science and technology policies and the injection of R&D funds, the government has helped to build the foundations for successful semiconductor manufacturing in Taiwan. ITRI was responsible for the execution of this policy, from know-how research through to technology transfer; it helped to build up the local IC industry and so to develop Taiwan as one of the world's leading computer and information technology providers. This success story has been impressive and the factors behind its success have caused it to be a key source for analysis and reference – the Japanese Industrial Technology Institute visited ITRI in 1995, for example. This chapter will introduce the history of ITRI, its main characteristics, its development process in IC business and its interaction with hi-tech manufacturers. It will then illustrate ITRI's role in the development of Taiwan's hi-tech industry, especially the IC industry, and show how it came to be seen as 'the cradle of technology' and the leader of the industry, and how it views its current role and future plans.[1]

STATUS OF THE ORGANISATION: ITRI'S PAST AND FUTURE

July 2000 was ITRI's twenty-seventh anniversary. Looking back over its history, from the initial set-up and development stages to its current position, it is obvious that each decision-making turning point was a key influence in the development of Taiwan's IC industry, and played a vital part in its development.

In the early 1970s the Taiwan government decided to take charge of the development of the semiconductor industry by utilising the state's resources. The Minister of Economic Affairs, Mr Y.S. Sun, thought that the best strategic policy was to set up a research institute positioned as a not-for-profit foundation. The research could therefore be related to the needs of the country's industrial technology development, and hence achieve the dual aims of public interest and efficiency. At that time the three existing research centres under the Ministry of Economic Affairs (MOEA) – the Uni-industrial Research Centre (currently the Union Chemical Laboratories), the Mineral Industrial Centre (currently the Energy & Resources Laboratories) and the Metal Research Centre (currently the Mechanical Industrial Research Laboratories) – were joined together and relocated to Hsinchu. After the passage of the necessary legislation ITRI was born in 1973.

Initially ITRI consisted of only three divisions, with over 400 employees. Before it could begin to generate any revenues, employees' remuneration and development and other operating expenditures were funded by a government budget of NT$213 million. Over the next 20 years, because of the demands of domestic industrial technology development, ITRI progressively expanded and refined its organisation. For example, the Electronics Industry Research & Development Centre (now the Electronics Research & Service Organization, ERSO), set up in 1974, had a direct influence on the IC industry in terms of technology import and transfer and human resources development. The Computer and Communications Research Laboratories, set up in 1990, and the Biomedical Engineering Center also set up in 1990 (the history of ITRI's development is shown in Figure 2.1) were also important. In 2000, after the organisational restructure of the late 1990s (see below), ITRI consisted of seven divisions and four centres which were positioned as professional R&D business entities, and three service centres which were responsible for resource consolidation and allocation. There were also several R&D task forces organised across divisions or centres for mission consolidation and coordination (ITRI's organisation structure is shown in Figure 2.2). Employees total some 6100, and over half of them possess masters and doctoral degrees.[2] ITRI's annual revenue has increased from an initial NT$200 million to the current NT$17 billion. The government's budget resource has decreased from 90 per cent to 50 per cent, that is 50 per cent of ITRI's revenue (NT$8 billion) is rendered from undertaking government science and technology projects, and the other half is earned from industrial services. The government terminated financial support in 1994. ITRI only had eight patents in the first five years, and had accumulated only 53 patents by 1983, but it progressed rapidly in its second decade, accumulating over 1300 patents; between 1995 and 2000, more than

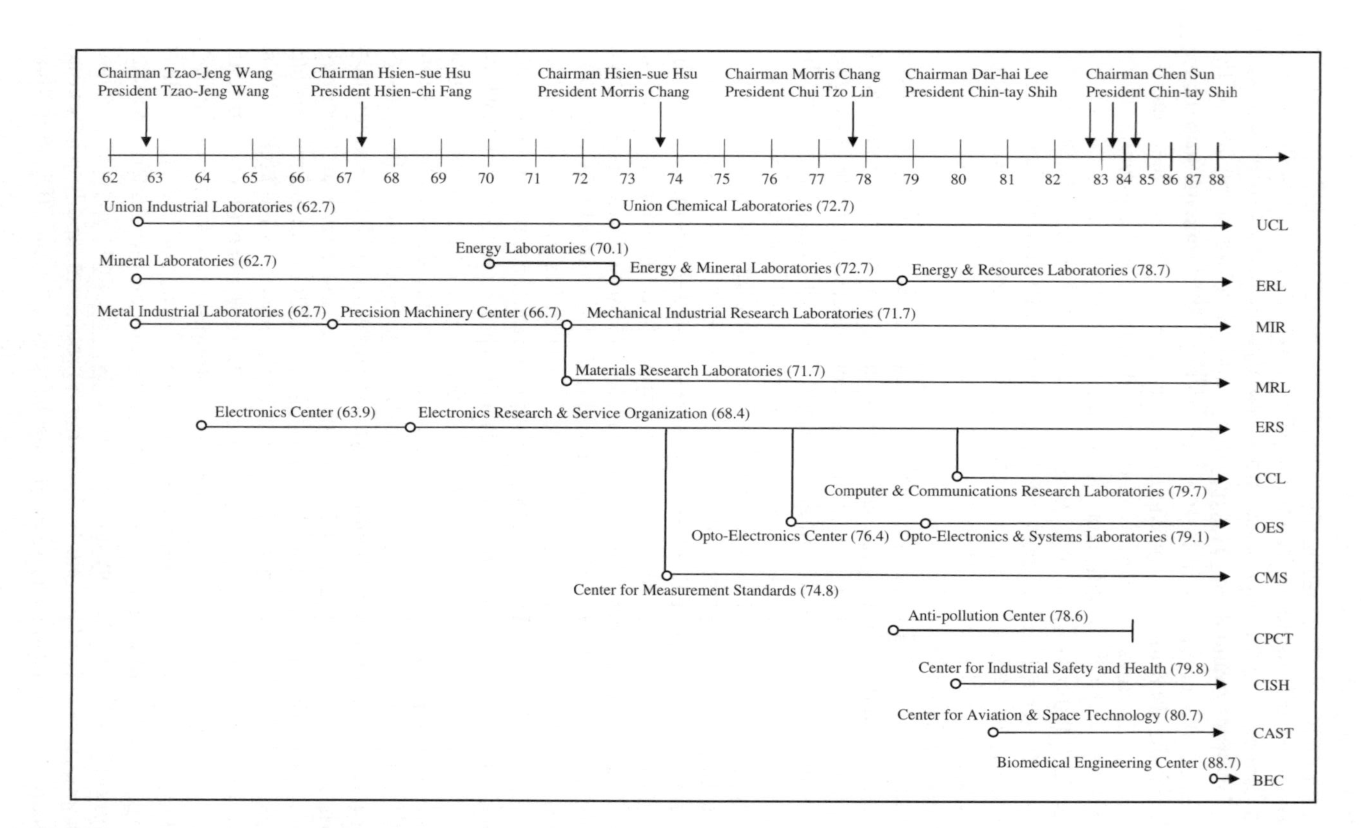

Figure 2.1 ITRI's historical development

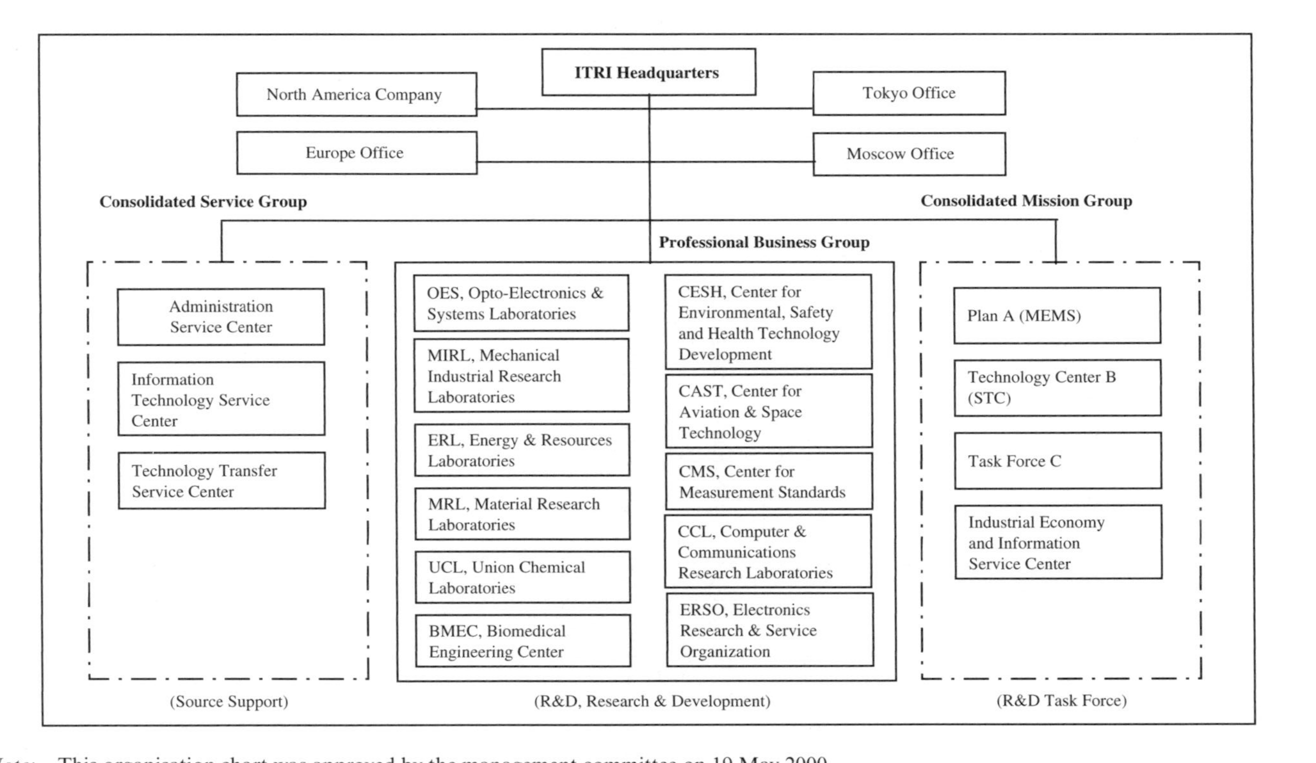

Note: This organisation chart was approved by the management committee on 19 May 2000.

Figure 2.2 ITRI's organisational structure

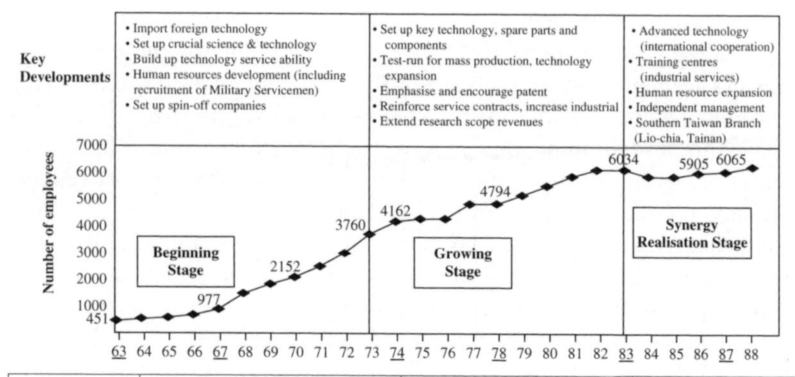

Function	Human Resources Development and Establishment of Independent Technology		Initiate and Assist the Industries		Industrial Partners and Main Resources	
Year	1963	1967	1974	1978	1983	1987
Total Revenue	240 million	576 million	4.45 billion	6.87 billion	13 billion	17.3 billion
Accumulated Patents	0	8	53	191	1366	3900
Divisions/Centres	3	3/1	5/1	5/3	7/4	7/3
R&D/Service	–	10:1	2.5:1	2:1	1.65:1	1:1

Figure 2.3 ITRI's progression & development – organisational structure (2000)

3000 patents were achieved. The total accumulated patents are now over 4000, and 53 are approved in the USA. ITRI has passed through its initial, growth and development stages, and its efficacy is universally appreciated (see Figure 2.3).

ITRI'S CHARACTERISTIC POSITION: 'EXIST FOR THE INDUSTRIES'

ITRI was set up at the very beginning with a special position – in simple terms it was to 'exist for the industries', and this is one of the main factors that has made ITRI so successful. As we have seen, ITRI was set up by legislation as a not-for-profit organisation supported by the government. There are only two similar types of institution in Taiwan, the National Health Research Institutes and the Chung-hua Institution for Economic Research. ITRI can be categorised as the country-level laboratory, reporting to MOEA instead of the National Science Council which is in charge of the R&D of the country's science and technology. ITRI is the MOEA's main executive arm for all science and technology projects, undertaking several billion dollars' worth of projects annually for the government. Compared with most other countries' national laboratories, which are focused more on basic research and academic research, ITRI's position is quite special, since it was always intended to perform *applied* research. In the early 1970s Taiwan's academic research was generally more fundamental and difficult to be commercialised, and the industrial sectors' research functions were engaged in product improvement and technology development and implementation, with little motivation for R&D. The gap between basic research and mass production had to be closed. That was the reason that ITRI positioned itself to do applied research from the very beginning – acting as the bridge between academic institutions and corporate organisations and concentrating on R&D of commercial value for industrial technology. It was also a mediator between the government and the private sector, executing the government's science and technology policies, physically responsible for the promotion of technology development and outcomes and participating in planning and advice in the formulation of government policy. ITRI thus played a key role in industry, government and academe.

ITRI's historic mission thus consisted of five fundamental elements: (1) to engage in applied research with emphasis on industrial efficiency and to rapidly improve industrial technology; (2) to develop flexible, commercial and advanced technology; (3) to diffuse the research results and technology to the industry; (4) in compliance with government policy, to assist SMEs to upgrade their technology; and (5) to cultivate industrial technology specialists. To successfully achieve this designated position it was necessary to adopt a thorough and comprehensive strategic management system. Concerning technology R&D, in order to meet industrial requirements and upgrade the industrial level, ITRI generally adopted a 'pull-push' strategy, trying to find a balance between market demand pull and technology development. While the market demand existed, in order to meet the demand gap in the short term ITRI's mission was to coordinate the industries' requirements and initiate technology R&D in product development, process improvement, etc. In the longer term it needed to be ahead of the industry by engaging in R&D in innovative and advanced technology, and to create new industrial fields. As ITRI was positioned for applied research, it also had to test run mass production for the R&D industrial technologies, in order to ensure that such technologies were feasible. At the same time, it also had to plan how to strategically reduce its role once the industry's required technology level had been achieved. ITRI's position could be described as solving the following problem: 'If the technology cannot be promoted, that is because the R&D technology is not good; but if it is successfully developed and becomes mass production, then it cannot be sold next year.' As a not-for-profit organisation ITRI had also to adhere to fair and open dealings, and regularly present its technology and R&D results to the industry: ITRI therefore never considered release of a shareholding to its employees.

ITRI spent a great deal of effort on human resources development. In addition to recruiting overseas specialists returning to Taiwan to work, it very soon set up a human resources development programme that allowed employees to continue to pursue their doctoral degree or study management courses at either Stanford or MIT, both universities that were clearly positioned to promote industrial technology R&D. In the early days ITRI had only some ten employees holding a doctoral degree, and it sent its employees to the USA at an average cost per head of some NT$5 million. Among the current 6100 employees there are 871 possessors of a doctoral degree, 120 of them sponsored by ITRI. Thanks to this human resources development system, ITRI gradually built up its own technology, and this human resources advantage gives it an edge over other Asian countries like Hong Kong and Singapore who are also aggressively developing their hi-tech industry. In order to maintain a pool of innovation ability ITRI sets its employee turnover rate at 10–15 per cent, in the hope that new blood will bring in new concepts and technology and so keep ITRI ahead in the industry. Mature technology can then be spread through industry when the well-developed specialists join private firms. Currently ITRI recruits on average 100 doctoral degree employees to replace 90 resigning employees every year, and the turnover of undergraduates and graduates is around 400–500. The average annual turnover is around 800–900 employees, and each employee's average seniority is about five years. As the accumulation of technology depends on the retention of employees, ITRI also benefits from the government's 'national defence military service' policy. Especially in the initial stages, when ITRI encountered difficulties in recruitment and a human capital gap, those employees undergoing this type of military service were crucial to ITRI's manpower supply. In the late 1980s Taiwan's stock market was booming, the private industries' compensation level was several times ITRI's, and many employees from ERSO and the Computer and Communications Research Laboratories were being headhunted by the private sector; the turnover rate reached almost 30 per cent, some departments suffered the loss of over 20 staff at the same time and the majority of the remaining employees were the military services staff who stabilised and maintained technology standards. Currently the Ministry of National Defence assigns 400 national military service soldiers to ITRI every year, and 70 per cent of these are undergraduates or graduates of the top-notch universities in Taiwan (National Taiwan University, Ching Hwa University and Chiao Tung University). The selection rate is very competitive as only 10 per cent of national service soldiers are qualified for this type of job, and they are obliged to stay for at least four years (previously six years): every year ITRI can thus expect to receive an outstanding new injection to the workforce.

ITRI is thus able to play a key catalyst role in the industrial development process, and to became the technology centre of the industry. Through recruiting good quality new employees, internal training and development, and maintaining adequate employee turnover, the specialists cultivated by ITRI are eventually transferred and contribute to private industry. This role and contribution are clearly illustrated by the way in which ITRI promoted and supported the birth of Taiwan's IC industry.

THE BIRTH OF THE IC INDUSTRY: THE FOUR PHASES OF ERSO'S PROJECT

To assist Taiwan to develop an IC industry, after setting up ITRI MOEA set up the Electronics Industry Research & Development Center (the predecessor of ERSO), to take charge of IC technology transfer and development, and it launched the project in four phases. Starting from scratch, ITRI progressively raised Taiwan's IC technology to a world-class level.

Phase I: IC Demonstration Plant Set-up Plan (1975–79)

When the Taiwan government decided to choose the electronics industry as the 'strategic industry' the country's lack of independent IC R&D capability meant that the know-how transfer of IC technology was dependent on the advanced countries. With the assistance of the 'Technology Advisory Council' (TAC), a group of overseas Chinese organised by Wen-yuan Pan, ITRI actively invited the USA to join in technology cooperation, with the proviso that the guidelines for selecting technology must fit the industrial characteristics of Taiwan's smaller but flexible business scale, and also possess the potential to be rapidly commercial. After some period of consideration, RCA was selected as the cooperation partner, and a technology transfer and authorisation contract was signed in 1976. Through the manufacture of chips for electronic watches the processing know-how of 7μm Complementary Metal Oxide Semi-conductor (CMOS) was imported, and an IC demonstration plant set up to test run in 1977. Four months after the plant was set up it had achieved the mutually agreed success rate (80 per cent), and later it even exceeded the RCA technology centre's own standard. In 1978 the accumulated production output for the chip was over one million sets (Sze-hwa Wu and Lung-chin Sen, 1999). This was the successful prelude to Taiwan's IC development.

The successful experience of the IC demonstration plant gave ERSO a firm foundation in IC manufacturing and design. But IC industry competence relies on constant investment and expansion of production, and these factors are not affordable for a not-for-profit organisation such as ITRI. As the country's research institute, ITRI's mission was the transfer of IC technology to the private sector; if foreign companies set up plants in Taiwan, the existing human resources would be attracted to those companies and would undermine the domestic IC industry development (Li-ying Su, 1994). The IC technology would then have to transfer to the private entities who would carry out the mass-production technology, and ERSO would continuously develop the upstream and downstream know-how. At the suggestion of the Division Head, Ding-hwa Hu, and the demonstration plant's Plant Manager, Chin-tay Shih, MOEA decided to set up the first *spin-off company*, a privately owned entity to take over the results of technology development and transmit it to industry. In order to privatise the company MOEA invited local conglomerates to participate in the investment. After overcoming several obstacles, such as the attitude that 'the more understanding there is about the IC industry, the less courage there is to invest', the first privately owned IC manufacturer, UMC (see Chapter 6), was finally born.

Unlike normal technology transfer, ERSO from September 1979 made a total plant operation transfer to UMC. UMC was officially set up in 1980, and until 1982 worked on completing the test run; ERSO's technology transfer project team members were heavily involved in support and assistance, including plant design, equipment specification and layout, machinery installation and test runs, and staff training. ERSO also sold UMC the existing know-how, self-designed products and work-in-progress research plans at a very low price. UMC's key management and technical employees were also transferred from ERSO. The Deputy Chief of ERSO, Hsin-cheng Tsao, was promoted to be UMC General Manager, and ERSO's Testing Department Manager, Yin-dar Liu, and Marketing Department Manager, Ming-tze Hsuin, became UMC's Assistant General Managers (Sze-hwa Wu and Lung-chin Sen, 1999). The whole plant technology transfer was unique, thorough and successful. UMC's 4-inch wafer plant started mass production in April 1982, and broke even in November of the same year. In 1985 UMC's profitability was ranked number one in the country's top 500 private companies. This caused great excitement in ITRI, who dreamed of independently developing an IC industry, and among Taiwan government science and technology officers.

Phase II: Electronics Industry R&D Plan (1979–83)

With the foundation laid in phase I for continuously developing existing manufacturing know-how and reinforcing research into large-scale and high-density IC automatic design technology, ERSO subsequently began the four-year phase II in July 1979. At the same time, in order to improve domestic IC product development, ERSO imported technology from the International Material Research Company (IMR) in the USA, whose technology level was lower but could satisfy domestic demand for optical mask duplication technology (Sze-hwa Wu and Lung-chin Sen, 1999). From July 1977, the time of signing the technology transfer contract, to April 1979, the time of setting up the Mask Department, ERSO officially provided the mask outsourcing services for Taiwan IC manufacturers and in 1981 became the only institution possessing the mask duplication know-how, constantly performing very well in terms of just in time (JIT) products and output. Until 1988, because of the large increase in IC design houses and manufacturers, the demand for mask technology was very great, and some foreign companies intended to invest in mask plants in Taiwan. In order to retain local specialists, ITRI again decided to set up another spin-off company, and invited ERSO's Mask Department Manager, Bi-wan Chen, to be the new company's General Manager to organise the company and transfer the technology. In October 1988 the Taiwan Mask was officially set up.

Phase III: VLSI Technology Development Plan (1983–88)

After the completion of the phase II plan, with NT$2.9 billion government support, ERSO initiated the 'VLSI technology development plan' aimed at developing the automated design capability for VLSI, becoming the technical support centre for the Taiwan IC industry, and providing the core components for the electronic equipment. In order to select the technology ITRI decided to build on the foundations of the first two phases and to develop design-intensive products more suitable for Taiwan's industrial characteristics, rather than developing in the technological direction of mass production and standardisation. ITRI also decided not to directly transfer the know-how from large foreign manufacturers but in 1983 chose to work with Mosel Vitelic, a company set up by overseas Chinese in the USA. Both parties signed technology cooperation and product development contracts for jointly developing the 1.5μm CMOS know-how, and designing and manufacturing 64K–256K DRAM (Kuan-fu Chen, 1990). This technology development was successfully achieved in 1985, and a design centre was set up to provide domestic manufacturers with the IC product design-aid tools and services.

Although there was substantial improvement on the VLSI design technology, the capacity of both ERSO's and UMC's manufacturing equipment could provide only the production of LSI, so Mosel Vitelic sold its developed technology to Korea. Later some companies like Mosel Vitelic sold similar technology or outsourced production to Korean and Japanese manufacturers (Kuan-fu Chen, 1990). That made the government realise the urgency of building up the VLSI industry. But because the number of fabless IC design houses was increasing, the ITRI President, Morris Chang, thought that the professional IC foundry had great strategic potential, and was also very suitable for Taiwan to develop as it possessed very good processing capability. In the light of that strategic viewpoint, the government decided to set up another independent privately owned company, Taiwan Semiconductor Manufacturing Company (TSMC, see Chapter 5), to industrialise the research results of the VLSI development plan. Even though there was the successful example of UMC, fund raising for TSMC was not as smooth as expected, as a majority of domestic conglomerates still hesitated. Formosa Plastics Corporation, for example, initially agreed to invest, but then hastily sold out its small shareholdings (Der-lin Fang, 1994). The joint venture was eventually concluded when Philips agreed to invest, planning to utilise TSMC as its plant in the Far East to compete with Japan and Korea (Kuan-fu Chen, 1990). Morris Chang, however, negotiated with Philips to make TSMC an independent company, rather than a Philips' satellite plant (Ai-li Yang, 1997). In 1986 ERSO first set up a 6-inch VLSI preparatory plant, and in October arranged the full transfer plan. In February 1987 TSMC was officially set up; because of the meticulously planned strategy and execution, TSMC's revenue was NT$2 billion over the first two years, and has since become the number one professional IC foundry, another successful case of an ITRI spin-off company.

Phase IV: Five-year Plan for Developing Submicron Processing Technology (1990–95)

After the completion of the phase III plan, ERSO launched the 'four-year plan for microelectronic technology development' to upgrade the VLSI design technology, but this plan was not successful. So, from July 1990, in order to coordinate with the government's science and technology development plan, ERSO started to operate a 'five-year plan for developing submicron processing technology' in order to build up the country's VLSI R&D and manufacturing capability, and also utilise a spin-off company to develop the know-how for 8-inch wafer mass production. However this plan provoked the IC manufacturers' suspicions and objections, particularly concerning the impact of the new spin-off company on existing firms. After several rounds of discussion, it was agreed that the government should cooperate with the private sector to develop the submicron plan. A 'Submicron Plan Consulting Committee', formed by the representatives of industry, government, academe and research, was responsible for consultation and coordination. A 'Submicron Association' was formed by UMC and TSMC under the project plan, to arrange staff exchanges to learn and cooperate and to obtain the primary technology authorisation. There were another six companies organised in a 'Submicron User Association', who assisted new product development and helped to test run technology within the shortest possible period. Among those companies, Etron Technology, Inc. was responsible for product design and staff training.

Compared to other countries' investment in submicron R&D, the budget for the plan was quite tight (NT$7 billion for five years): the annual R&D budget of Korea's Samsung was more than the total amount of Taiwan's science and technology project budget. However, the government's financial support and joint efforts between ITRI and the private sector meant that the plan not only successfully built up the 0.35μm processing mould technology, which enabled Taiwan IC technology simultaneously to keep pace with international competitors, but it was also completed six months ahead of schedule, which saved an average of NT$1 million in interest per day for Taiwan (Jun-chung Chen, 1995). ITRI's fourth spin-off company, the Vanguard Semiconductor Company, was officially set up in September 1994. The leader of the 'submicron plan' and also ERSO's Deputy Head, Tze-yuan Lu, was appointed the company's Assistant General Manager, responsible for leading the company from laboratory testing to mass production. Vanguard was not only Taiwan's first 8-inch wafer manufacturer, its DRAM manufacturing symbolising the Taiwan IC industry's move from historical non-standardised consumer IC products to standardised products, but also the only self-directed technology company in Taiwan, unlike the other DRAM manufacturers (Powerchip, etc.) whose technology was transferred from the USA and Japan.

Through successfully executing the government's four-phase science and technology projects, ITRI thus played a key role in Taiwan's IC industry development. In terms of technology development, ITRI rapidly pushed forward Taiwan's IC technology many generations within 20 years (1975–95) (see Figure 2.4). Under the phase I plan, in 1976, ITRI introduced 7μm processing know-how, which was far behind the current most advanced 3μm processing know-how; but in 1985, in the phase III plan, Taiwan's processing know-how was improved to 1.5μm, which was only three years behind the most advanced 1.2μm processing know-how. In the submicron plan, ITRI's 0.35μm processing know-how was ahead of the world production line and was almost comparable with the USA and Japan. For the IC industry, each jump in technology generation means that Taiwan possesses a greater cost advantage, because on the same size of wafer the smaller the path, the higher the unit capacity. Comparing the most advanced 0.15μm know-how with 0.25μm know-how, for example, it

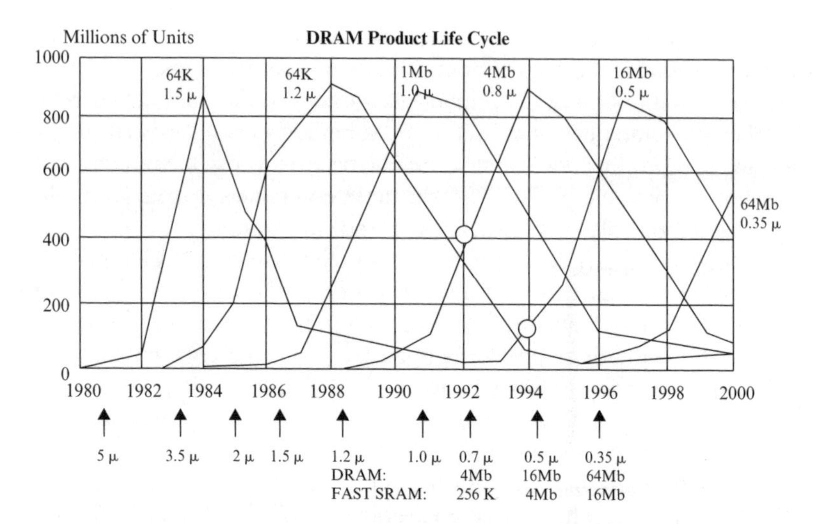

Note: The curve represents the output and time of the world IC (DRAM) processing technology; the figures shown on the bottom and the arrows show the time when Taiwan participated in each stage of processing technology.

Figure 2.4 Comparison of the IC industrial development for Taiwan and the world

can produce an additional 116 ICs on the same size of wafer. That also explains how Taiwan's IC industry can so rapidly enter into the manufacture of a new product and generate additional profits. As Taiwan's IC industrial technology reaches the international standard, the more easy it is for Taiwan's manufacturers to obtain cooperation with other countries, and the room for price negotiation is substantially increased.

In terms of industrial development, through the spin-off companies set up by ITRI at each phase, the mature industrial technology and manpower were seamlessly transmitted to the private sector, but the benchmark functions for the prosperous development of Taiwan's IC industry were also set up (see Figure 2.5). The phase I IC demonstration plant created UMC, which became the symbol of Taiwan's entry into the IC industry, and also stimulated the establishment of IC design companies (Syntek Semiconductor Ltd and Holtek Semiconductor Corp., etc.). The phase II Taiwan Mask provided Taiwan manufacturers with self-made mask services which completed the value chain between IC design and manufacturing. After the phase III VLSI plan had created TSMC, UMC and TSMC successfully encouraged local companies to invest, and many companies were set up (such as Hualon Microelectronics and Winbond). The wafer

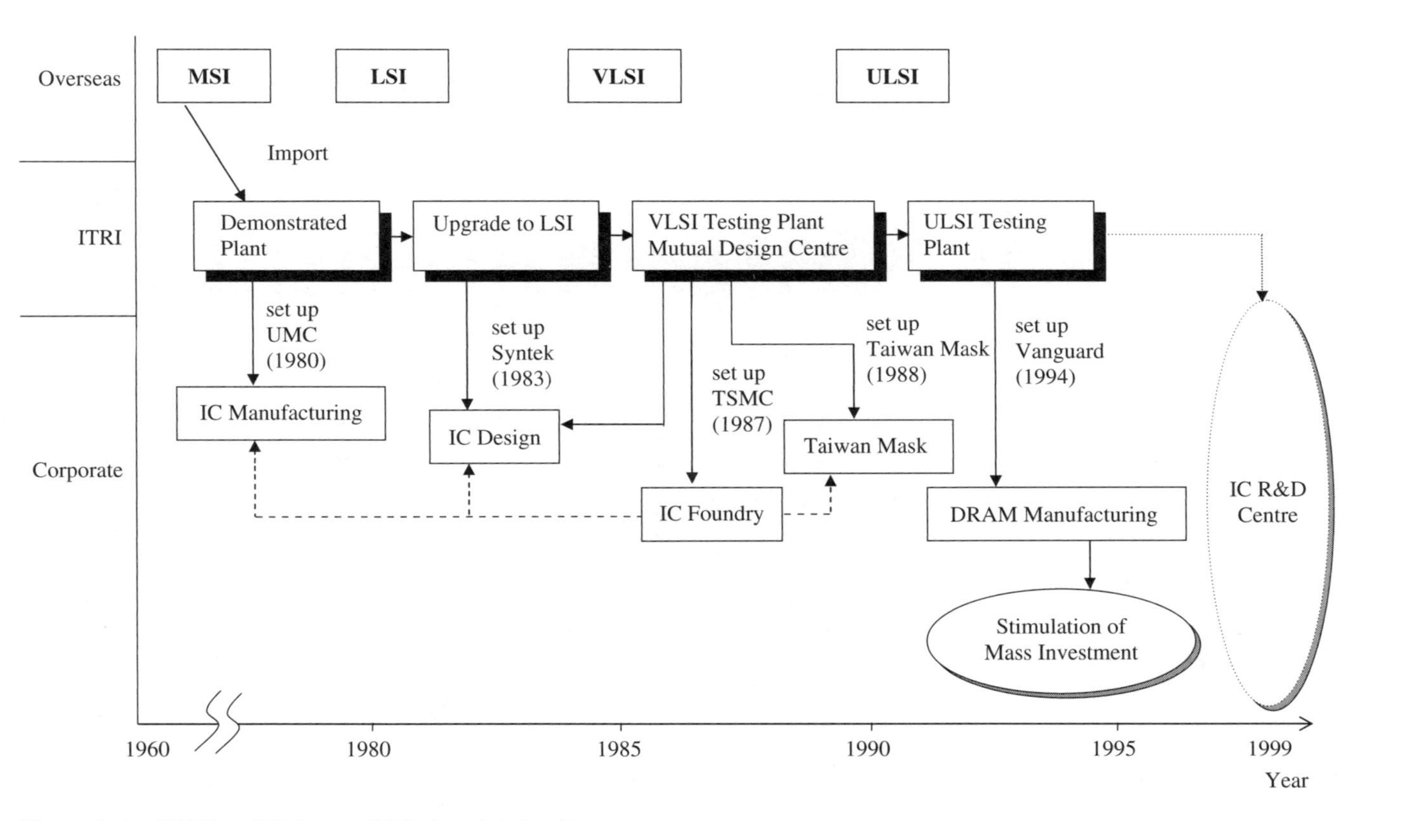

Figure 2.5 ITRI and Taiwan's IC industrial development

foundry and joint design centre directly promoted the establishment of many IC design companies. The submicron plan successfully created Vanguard, and induced an investment boom in Taiwan's 8-inch wafer and DRAM manufacture. Within a short period over NT$600 billion capital had been attracted and it was hoped to build 24 plants. Taiwan manufacturers began mass participation, ranging from wafer design to processing, testing/packaging, etc., creating a comprehensive production system and a mature Taiwan IC industry.

INTERACTION WITH HI-TECH MANUFACTURERS: THE NETWORK AND SUPPLY CENTRE FOR TECHNOLOGY AND SPECIALISTS

In addition to the contribution of upgrading IC technology and developing the industry as described above, ITRI was the network and supply centre for both technology and manpower within Taiwan's hi-tech industry. This can be illustrated by its long-standing cooperative relationship and technology transfer with high-tech manufacturers, and the history of its employees' career patterns.

Technology Transfer and Cooperation

In addition to each spin-off company directly taking over the research results from the projects in each phase, many of the private IC companies also obtained the relevant technology and knowledge from ITRI. Particularly before 1990, many private manufacturers were set up by transferring technology from ITRI – for example, the IC designers Syntek Semiconductor Ltd and Holtek Semiconductor Corp. (1983), Proton (1985), and then the Silicon Integrated System Company (1985) all accepted the IC design spec transfer. Among IC manufacturers Advanced Device Technology Inc. (ADT) (1986), Hualon Microelectronics and Winbond (1987) obtained the sources of technology from ITRI (see Table 2.1). After 1990, because of the improvement in manufacturers' skills, the percentage of joint developments with ITRI increased. In 1989, ERSO developed the VGA with Winbond, and in 1993 it developed the 50V CMOS FET with Holtek Semiconductor Corp. The submicron plan started in 1990 involved several existing IC firms in joint research, development and cooperation. Therefore it can be said that the Taiwan IC industry relied upon ITRI as a technology supply centre and production network. The current interaction between the ITRI and the IC industry is in fact not just focused on ERSO. In addition to ERSO, engaged in the

Table 2.1 ITRI's technology transfer to the IC industry

Date	IC design technology		Date	IC manufacturing technology	
	Transfer target	Content		Transfer target	Content
1983.11	Syntek	LOVAG Design Specification	1979–1982	UMC	Transfer of Whole Plant Operation and Management Specialists
1983.11	Holtek	LOVAG Design Specification	1986	ADT	LOVAG CMOS Process
1984.3	UMC	TOCMOS Design Specification 3.5 μm SACMOS Design Specification	1986–1987	TSMC	Transfer of Whole Plant Operation and Management Specialists
1985.1	Syntek	7V SACMOS Design Specification	1987	Hualon	Bipolar Process
1985.6		Bipolar-SS Process			
1986.3		5V LOVAG Design Specification			
1985.9	Holtek	5V LOVAG Design Specification TOCMOS Design Specification			
1985.9	Proton	Bipolar-SS Design Specification Bipolar-SD BiMOS Design Specification Bipolar-I^2L Design Specification			
1985.11		LOVAG Design Specification TOCMOS Design Specification			
1988.7	SIS	5V LOVAG Design Specification			

Note: The data are summarised from ITRI's 'Follow-up Analysis of the Influence on the Industries Caused by the Semiconductor Project', pp. 37–8, 1987, and other internal data.

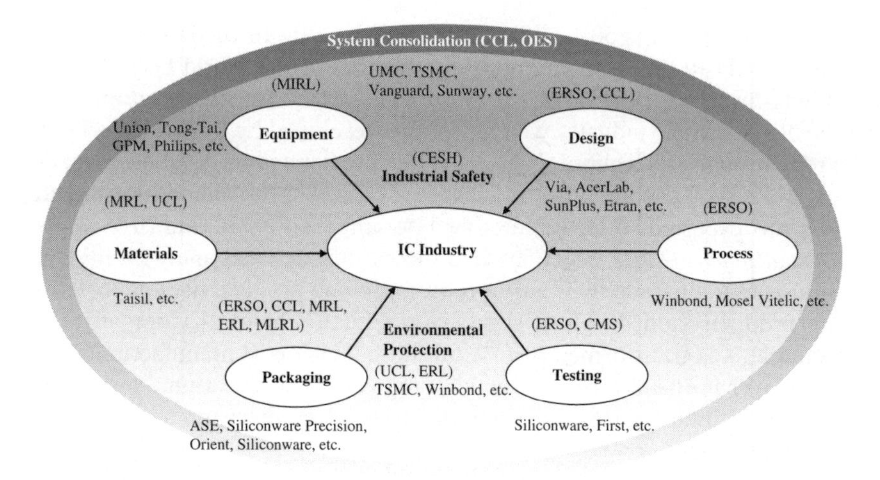

Notes:
() ITRI's Divisions/Centres Involved.
The companies listed are those who accepted the technology transfer or cooperative research.

Figure 2.6 ITRI's relationship with Taiwan's IC industry

development and transfer of process technology, there are the Mechanical Industrial Research Laboratories engaged in the R&D of semiconductor equipment, and the Materials Research Laboratories and Union Chemical Laboratories developing the materials for wafer. The Center for Measurement Standards concentrates on research in testing technology. The Computer and Communications Research Laboratories and Opto-Electronics & Systems Laboratories (OES) develop the technology for consolidating design and systems, and they all establish a technology transfer and cooperative research relationship with the IC firms (see Figure 2.6).

We can also see ITRI's efforts being expended on the evolution of other new hi-tech industries. Before 1994, for example, there was no opto-electronics industry in Taiwan; as ITRI's OES successfully accomplished the R&D of epitaxy materials the technology was transferred through the setting up of Highlight Optoelectronics Inc. Taiwan's opto-electronics industry thus started from zero. From the original role of assembly of CD-ROMs for foreign brands to self-developed DVD, 75 per cent of Taiwan's opto-electronics companies worked with ITRI for technology transfer and cooperation, a total investment of over NT$10 billion that constructed a complete industrial structure. The total output of CD-ROMs in 1998 was 34 million sets, accounting for 38 per cent of world volume. The estimation of 2002 output was 50.1 million sets, making Taiwan the manufacturing king of CD-ROMs, and an important member of the World DVD Forum (Gung-hwa Lin, 1999; *ITRI Annual Report 2001*). Another example was the display writer industry which was also developed by ITRI's know-how breakthrough and technology transfer that accelerated the establishment of many TFT-LCD companies and involved over NT$100 billion investment: this was expected to create another booming Taiwan IC industry (*ITRI Annual Report, 2001*). ITRI has also successfully assisted many traditional industries to upgrade their business by improving productivity and increasing products' added value. The Chang Chuan Chemical Company, for example, was transformed from a traditional chemical manufacturer to a key player in the electronic chemicals industry. The traditional industrial machinery industry was also upgraded to produce precision and automation machinery.

Spread and Influence of Professional Personnel

The impact caused by the dispersal of ITRI employees within the Taiwan hi-tech industry was almost more crucial than the effect of its direct technology transfer and cooperation policies, since the contribution of such professional personnel could be broader and more durable. As mentioned earlier, ITRI has consistently cultivated its employees, and when its mature professional employees left, the majority of them continued their career by joining the relevant industry. According to ITRI's records, from 1973 until the end of May 2000 a total of 13995 ITRI employees moved to other fields, with 11065 (about 85 per cent) joining industry, and only 15 per cent joining academe and government: this is quite different from the situation in the USA and Japan. Among those employees who switched to industry, almost a quarter (24 per cent) worked in the electronics and IC industries, next was information and communication (19 per cent), and then machinery and automation (16 per cent); others worked in opto-electronics and materials. This allocation is very close to the profile of the recent development of science and technology in Taiwan. Among the 11000 ITRI ex-employees in industry, 4464 worked at the Hsinchu Science-based Industrial Park (HSIP, see Chapter 3), and most of them were spread over the IC, computer and peripherals industries.

Along with the impact of these professional personnel, ITRI's influence on the hi-tech industries has steadily become more profound. Currently 82000 people work at HSIP and 12000 (15 per cent) of them are at management levels, 1.2 per cent possessing doctoral degrees and 12 per cent masters degrees. Over 50 per cent of ITRI employees possess masters and doctoral degrees; once they leave to join industry, 80–90 per cent of them will take up management positions. That also means that about 40 per cent of the managers in HSIP will have worked for ITRI. Observation of the IC industry's executive lists will show many senior personnel who have contributed to the development of Taiwan's IC industry holding decision-making positions in each company (see Table 2.2). The hi-tech companies therefore not only took over ITRI's technology and knowledge, but also its employees' management experience. For example, the management organisation of TSMC and Taiwan Mask is exactly the same as ITRI's, which saves managers a lot of learning time (Ming-shuen Lee, 2000). Morris Chang formulated the strategy for TSMC, and the professional IC foundry has become the prototype for its global competitors. Ding-yuan Yang, the Director of ITRI's Planning Department, assisted in the setting up of Winbond and suggested the employee stock bonus policy, a policy which has become the benchmark for hi-tech human resources in the industry. As those managers all possess an ITRI background, and even graduated from the same universities (Chiao Tung and Ching Hwa, etc.), their social network has gradually became the information source, interaction channel and foundation for the companies' cooperation. Many years ago the personnel managers of UMC and TSMC organised a club for exchanging information and developing common agreement on salaries, compensation and recruitment policies. In 1996 more than ten IC manufacturers organised an association to jointly conduct international recruitment, deal with foreign large IC companies' law suits for patent penalties, and set up a mutually owned patent company. This intensive network of interpersonal relationships greatly contributes to the frequent interaction, close cooperation and transparent information sharing among the companies in HSIP.

CHALLENGE AND PROSPECTS: ITRI'S CHANGING ORGANISATION

Today the overall environment of Taiwan's hi-tech industry has substantially changed. The challenge of fierce global competition and the private sector's quality and capacity was getting stronger, while ITRI, the historical industrial vanguard organisation, was also getting larger. It appeared that the original mutual supportive relation between ITRI and the industries was gradually being changed as competition and conflict grew. Faced with environmental challenge and change, ITRI decided to reorganise entirely, striving to recreate its industrial interests and restore its previous success.

Table 2.2 IC industry executives with ITRI background

Name	Position in ITRI	Current position	Remark
Morris Chang	Chairman	Chairman, TSMC	Spin-off Company
Fan-Cheng Tseng	Plant Manager	General Manager, TSMC	Spin-off Company
Hsin-Cheng Tsa	Deputy Head of Division	Chairman, UMC	Spin-off Company
Ming-Tze Hsuin	Manager	CEO, UMC	Spin-off Company
Yin-Dar Liu	Manager	Executive Director, UMC	Spin-off Company
Chin-Rung Hsu	Manager	General Manager, UICC	UMC's Subsidiary
Bi-Wan Chen	Manager	General Manager, Taiwan Mask	Spin-off Company
Ding Hwa Hu	Vice-President	Chairman, Macronix	
Ding-Yuan Yang	Director	Deputy Chairman, Winbond	
Ching-Jiu Chang	Head of Division	General Manager, Winbond	
Tse-Yuang Lu	Deputy Head of Division	Chairman, SC	
Kuo-zao Wang	Manager	General Manager, Syntek	
Bing-Tien Wu	Head of Division	General Manaer, HC	
Chi-Yong Wu	Chief of Section	Chairman, Holtek	
Ming-Jie Tsai	Manager	Chairman, Novatek	
Si-Ming Lin	Supervisor	General Manager, WC	

Note: The data was updated on 10 August 1998.

COMPETITION AND CONFLICT

As already described, ITRI was the leading mover in technology at the earliest stage, and its relationship with the relevant industries was based on technology transfer and cooperation. In 1980, when UMC was set up, it requested ERSO to close the demonstration plant to avoid profit competition with the private company, but the existence of ERSO's demonstration plant had actually brought in positive competition, and its professional personnel were also UMC's major employee resource (Sze-hwa Wu and Long-chin Sen, 1999). But the objections of the private sector were much stronger when the submicron plan was launched in 1990. The controversial point was above all the establishment of spin-off companies. Basically the way the technology transferred to the spin-off companies was a type of cheap solution to switch high-class technology and specialists, as the investment and costs of the technology were subsidised by the government, helping the private sector to obtain the relevant know-how and human resources at a very low cost. If the spin-off companies participated in market operations with the normal enterprises, while enjoying the advantages of the government's subsidy and advanced technology, and their products also competed with those of existing manufacturers, the playing field for competition was obviously not level. The establishment of the spin-off company therefore required very critical assessment and needed to meet many strict terms and conditions – the product to be developed is not present in the existing market, and the technology to be researched must be one that the current manufacturers could not or did not want to develop. When ITRI set up Vanguard, for example, the product was limited to DRAM. However, UMC's Chairman, Hsin-cheng Tsao, also strongly objected that the submicron plan should not set up a spin-off company: 'The new company's product will definitely create a conflict with the existing manufacturers, or if it utilises DRAM know-how to make different products, the existing manufacturers will still not be able to compete with it' (Wen-heng Chao, 1993). It seemed that the private sector had begun to question ITRI's role; after the submicron plan was completed, and ITRI wished to develop a bigger-scale 'deep-submicron plan' it encountered very strong objections from the industry, and the budget was cut to NT$1.9 billion by the Legislative Yuan to obstruct any ITRI plan to set up another spin-off company. The industry also doubted the necessity for this plan, since at that time the R&D progress of many of the semiconductor manufacturers surpassed that of ITRI, and the government had no need to spend huge sums to conduct duplicate research (Sze-hwa Wu and Long-chin Sen, 1999). ITRI's early-stage R&D used to be regarded as too advanced and too risky, and few companies wanted to undertake the research results;

ITRI was now being criticised as too slow, and facing serious challenges from the private sector.

As the organisation grew bigger, the private sector's questions and criticisms increased – if the organisation grew too fast and it undertook a huge amount of hi-tech spending, could the budget be abused or efficiency undermined? Because ITRI exclusively led the country's industrial R&D, would that be good or bad for industrial development as a whole? Alongside a change in the industrial environment, the Taiwan government's policy also changed direction. In May 1993 ITRI's budget was substantially cut by the Legislative Yuan's annual budget review, and the MOEA also decided to cancel the regular NT$200 million annual budget listed in the terms and conditions for ITRI's creation (Wen-heng Chao, 1993). These changes of policy highlighted the expectations of government and private sector that ITRI, as a research institute, should reform and upgrade its position.

ITRI'S CHANGING ORGANISATION

To deal with these challenges, ITRI's current President, Chin-tay Shih, adopted a 'big-bang' change to reconstruct ITRI's competence, by reorientating ITRI's position and target, and entirely restructuring the organisation. Even though there was criticism from the private sector, the general expectation that ITRI would continue to play the role of 'cradle of technology' remained unchanged. It was simply assumed that ITRI needed to upgrade itself and engage in strategic R&D, to select the more risky areas where the industry had not prevailed, to perform preliminary exploration for the industry, and to leave commercial activities to the private sector. Based upon its responsibility as continuous service provider, ITRI established its future prospects and targets, including: reinforcing the R&D of advanced technology, improving the organisation's vitality, strengthening its industrial services, and promoting internationalisation. After a year of preparation, planning and communication, ITRI's organisational restructuring officially began in July 1999.

ITRI's so-called 'innovative research' means small-scale research containing the potential for creativity; the 'targeted advanced development' relates to large-scale research with clear direction possessing long-term potential. In order to achieve the target of setting up a new hi-tech industry, ITRI expected to increase the ratio of innovative research and targeted advanced development to 40 per cent in 2000, expecting that considerable synergy would be realised in 5–10 years. The ultimate objective is to build up the country's leading technology and to divest the industry of a 'follower' position (Chin-tay Shih, 1999). To build up the innovative and advanced research mechanism, the organisation needs to improve its vitality, and the most important measure is to amend the organisational structure. Figure 2.2 illustrates the new organisation structure categorising the 12 professional research divisions/centres as five major fields which can jointly undertake the hi-tech projects, resolve internal competition and reinforce resource consolidation. This will also permit the flexibility to organise the task force to cope with the demands of technology development, and also strengthen cooperation across divisions and centres. To adopt the concept of sharing service centres, to simplify the internal process procedures and to improve efficiency, the administration and procurement functions are to be consolidated as one service centre. To strengthen the system of human resources training and development it is essential to set up a reward policy to motivate employees' innovative spirit. ITRI will expand the range of its laboratories open to the industry, and to provide them with the services of joint development and commercialisation. ITRI also offers entrepreneur support services, based on the consideration of customers' demands and the need to supply a 'one-stop service'. To coordinate the government's Tainan Science-based Industrial Park (TSIP) plan (see Chapter 3) and to increase the services provided to Southern Taiwan's customers, ITRI plans to set up a branch there, estimating its total employees to be about 2000 by 2008. Finally, in terms of promoting internationalisation, ITRI is not only assuming a mission to perform as the bridge linking industry, government and academe, but is also attempting to join up domestic and international technologies. As Taiwan's annual R&D budget accounts for less than 1 per cent of total world volume, it is very important to expedite R&D efficiency through international cooperation.

ITRI's re-engineering programme is still being carried forward aggressively. These changes include the total re-election of its ninth term Board of Directors, a major change that saw 13 new directors join. Several groups set up for different aspects of change and 170 employees are engaged in 15 types of re-engineering projects. It is anticipated that one-third of the employees will have to undergo adjustments to their job duties as a result of these changes. Without precedent, ITRI has set up four service centres in two years (Tze-fung Lu, 2000). The results of these changes have been increasingly apparent: since the laboratories opened in 1996 they have attracted many hi-tech companies and entrepreneurs with a total of more than 1500 users, and 38 new companies have been successfully assisted to set up. ITRI's training and development centres are the best among the 46 similar centres in Taiwan, and are regarded by the government as a successful instance of cooperation with the private sector. In 1999 the government passed the 'Elementary Law for Science and Technology' which permitted ITRI to enjoy partial results of its intellectual property rights, a law which should bring benefits for ITRI's research motivation and results applications.

Not simply ITRI itself, but all industry in Taiwan is expecting that the former 'cradle of technology' will successfully re-invent itself and become industry's future indispensable resource partner. The crisis could in fact turn out to be a turning point: whether or not ITRI takes this opportunity to pursue further success will depend on its current endeavours.

NOTES

1. The author very much appreciates the help of Mr Dar-hsien Lo, the Director of Planning, ITRI, who agreed to be interviewed. This chapter was written based on the interview contents and the information he provided.
2. ITRI statistics, unless otherwise specifically noted, are based on data from May 2000.

REFERENCES

Chao, Wen-heng (1993), 'Industrial tranche? Scientific emperor?' *Excellence*, July, 8–78.

Chen, Jun-chung (1995), 'Tze-yuan Lu – making the wafer by his will', *Common Wealth*, 165, February, 88–91.

Chen, Kuan-fu (1990), 'The development and space structure which Taiwan hi-tech depends on – the case of the Hsinchu Science-based Industrial Park', *Taiwan Social Research Quarterly*, **1**, 13–149.

Fang, Der-lin (1994), 'The conglomerates' march toward the hi-tech industry', *Wealth Magazine*, **150**, September, 232–6.

ITRI (2001), *Annual Report 2001*.

Lee, Ming-shuen (2000), 'Fierce competition in technology', *Common Wealth*, **228**, May, 124–34.

Lin, Gung-hwa (1999), 'A maturing opto-electronics industry', *Technology Pioneer*, March, 3–4.

Lu, Tze-fung (2000), 'ITRI and III (Institute for Information Industry) took the train of transformation', *Common Wealth*, **232**, September, 86–193.

Shih, Chin-tay (1999), 'In search of innovation and facing the cross century challenge – ITRI's innovative and advanced strategy', *Technology Pioneer*, July, 4–5.

Su, Li-ying (1994), *ITRI's 20 year History*, Hsinchu: ITRI, ERSO.

Wu, Sze-hwa and Sen, Lung-chin, (1999), 'The formation and development of Taiwan's semiconductor industry', *Taiwan Industrial Research*, **1**, 57–150.

Yang, Ai-li (1997), 'IC godfather – Morris Chang', *Common Wealth*, **190**, March, 116–40.

3. Science parks in Taiwan: HSIP and TSIP

Soo-Hung Terence Tsai and Chang-hui Zhou

TAIWAN, CHINA AND THE USA: HISTORICAL OVERVIEW

Despite a history of conflict and civil war (see Table 3.1), Taiwan occupies a crucial position in the triangular US–China–Taiwan geopolitical configuration, as the so-called 'hot button' between the USA and mainland China. Taiwan has historically been one of China's main sources of foreign direct investment (FDI) and until the mid-1990s bilateral relations were beginning to improve in other areas.

ECONOMIC DEVELOPMENT

Taiwan's Economic 'Vital Statistics'

Table 3.2 gives the 'vital statistics' of Taiwan as a focus for industrial development.

Table 3.1 Taiwan, China and the USA: historical background

Date	Events
16th century	Taiwan ceded to Japan
1945	Taiwan ceded by Japan to Communist China
1945–47	Taiwan–China Civil war
1946	Separation of Taiwan from Communist mainland China, neither recognising the legitimacy of the other
1971	Taiwan loses seat at the UN
1997–98	Greater Sino–US rapprochement under Clinton Administration
June 1998	Sino–US Summit in Beijing: Clinton declaration containing the 'Three No's'

Table 3.2 Taiwan: economic indicators, mid-1990s

Area	15 000 sq miles
Population	20 million
Language	Mandarin[a]
Law system	Basically European
Political infrastructure	Heavily US-influenced
Labour costs	Low
Economic growth[b]	6.42% per year
GNP *per capita*[c]	US$11 604
Industrial GDP[d]	US$90 billion
Foreign exchange reserves	US$80 billion
SMEs:	
as per cent of local firms	96
as per cent of industrial output	40
as per cent of exports	50
as per cent of employment	70

Notes:
(a) Amoy and Hakka are also spoken.
(b) Up to 1994.
(c) Up from US$200 in 1951.
(d) Up from US$260 million in 1951.

JOINING THE 'ASIAN TIGERS' CLUB

After 1950 Taiwan began to shed its past as a low-income and capital-deficient less developed country (LDC); government-promoted economic development initiatives sought to exploit the country's low cost advantages to establish private enterprise and overseas trade. In ten years the country had become a major export-oriented nation, based around the consumer electronics, plastics and textiles industries: Taiwan had become an 'Asian Tiger'. Table 3.3 shows the stages in this process, from import substitution, through export expansion, to the upstream raw materials industries of the 1980s.

GOVERNMENT INVOLVEMENT IN THE CREATION OF A WORLD-CLASS INNOVATION CENTRE

'Silicon Valley' in the USA is often used as evidence to show that many world-class innovation centres were formed without much government involvement. The case of Taiwan seems to challenge this argument. Taiwan was late in industrialising and there was a huge resource shortage during its early phase

Table 3.3 Taiwan: from import substitution to economic take-off, 1950s–1980s

Import substitution (1952–61)

- During the 1950s, the government's focus was on 'nurturing industry through agriculture and developing agriculture through industry'.
- Steady improvements in agricultural technology greatly boosted production and created export earnings to pay for the import of modern industrial machinery.
- Labour-intensive industries – textiles, plywood, foodstuffs and cement – rapidly developed and a solid foundation for continued economic development was laid down.

Export expansion (1962–71)

- During the 1960s, rapid economic growth and stable commodity prices were achieved by the adoption of export-orientation promotion strategies and government measures to encourage savings.
- The government also established export-processing zones (EPZs) to encourage exports and bring in foreign technology and capital; capital, energy and basic industries were placed on a sound development path.

Economic take-off (1972–80)

- Recession in the international economy and two oil crises in the 1970s meant that Taiwan's high dependence on trade and slow growth of exports constituted a brake on economic expansion.
- The government response was to institute major infrastructure projects and place more emphasis on the development of precision, upstream raw materials and heavy and chemical industries to serve as the foundation of further economic development.
- Taiwan's economic growth rate, which had bottomed out at just 1.16 per cent in 1974, recovered to 10 per cent in 1980.

Source: Based on an article in *Economic Affairs*, 1995.

of industrialisation. (Japan presented a similar case in the 1950s, when post-Second World War reconstruction was paramount: the Japanese government actively planned, reallocated resources, intervened and protected the infant industrial sector.) Following the so-called 'flying geese' model, Taiwan, like the other Asian Tigers, deliberately imitated the Japanese model. Government can be instrumental in four key ways, as shown in Table 3.4.

The Taiwanese government constantly steered the process from hard-currency accumulation to technology importation, to establishing a national laboratory (ITRI, see Chapter 2) that has spun off many high-tech firms to HSIP, to fostering environmental factors that facilitate industry clusters. What makes Taiwan unique is that the government never intervened in businesses at the micro level but developed a 'big-picture' blueprint for the entire

Table 3.4 Government intervention in economic development

- Facilitating resource reallocation and concentration for optimal exploitation
- Achieving economies of scale in utilising resources
- Maintaining focus, aiming high, and aligning business strategies and industry competition
- Promoting 'leapfrogging' to telescope development phases rather than adopt an incremental evolution model

island based on a vision and understanding of Taiwan's comparative advantages in worldwide competition.

APOC

The Asia-Pacific Operation Centre (APOC) plan best reflects how the government strategically positioned Taiwan to reap the benefits of the international division of labour. HSIP, UMC (Chapter 6) and Acer (Chapters 9–11) are so successful because there is such a clear understanding of Taiwan's comparative advantages.

US LINKAGES

Another key factor in Taiwan's success is its close links with the USA. Thousands of young Taiwanese scholars pursue their advanced studies at American universities and end up working in 'Silicon Valley', enhancing the linkages between technological development in the USA and Taiwan. At the beginning of the commoditisation of the PC in the early 1980s, it was the Taiwanese in 'Silicon Valley' who first spotted the trend, and recognised that Taiwan would be an ideal place to set up firms for business proliferation. Their efforts touched off the explosion of entrepreneurial activity that made Taiwan into the hinterland of 'Silicon Valley', a vital extension of the USA's high-tech industry.

THE MOVE TO HIGH-TECH DEVELOPMENT

Science-based Industrial Parks

In the later 1970s wage rates began to rise and the the New Taiwan dollar to appreciate, posing a serious challenge to Taiwan's economic development, as traditional labour-intensive industries began to lose their export competitiveness. The government's response was twofold:

- Promote trade and financial liberalisation
- Designate industries with strong market potential, advanced technologies and high value-added components for priority development

To attract high-tech personnel and industries, and to stimulate the development of industrial technology, a strategy of establishing *science-based industrial parks* was developed. The first science park was officially opened in Hsinchu on 15 December 1980: if developments in the 1970s had been a 'first take-off' for Taiwan, this might be considered the second.

SMALL AND MEDIUM-SIZED ENTERPRISES

Between 1980 and 1990 Taiwan changed from being a net capital importer to a major capital exporter. The products of basic industries and technology-intensive industries replaced those of traditional labour-intensive industries as the engines of industrial development. But Taiwan's small and medium-sized enterprises (SMEs) have been an equally important driving force behind the country's economic growth (see Table 3.2).

TAIWAN AS AN ASIA-PACIFIC REGIONAL MANUFACTURING CENTRE

'Intelligent' Industrial Parks

The government, anxious to safeguard Taiwan's leading position in the high-tech industry and its key position in the vertical and horizontal work-division system in the Asia-Pacific region, began after 1980 to develop the country as a regional operations/manufacturing centre. This strategy had the six elements outlined in Table 3.5, and was intended to break the bottleneck in industrial development. Five specific operations centres were to be set up, for manufacturing, transportation, finance, mass media and telecommunications. Developing Taiwan as a regional manufacturing centre was the cornerstone of the project.

The government planned to develop 20–30 'intelligent' industrial parks by the late 1990s to jointly produce high value-added and high-tech products, developed, operated and managed through a modern international telecommunications system.

Table 3.5 Taiwan as a regional operations/manufacturing centre: key target areas

- Promoting special projects for technological development and strengthening R&D
- Promoting the development of high-tech and high value-added products manufacturing industries
- Promoting the establishment of 'intelligent' industrial parks
- Promoting the development of EPZs and setting up special transshipment and warehousing zones
- Actively promoting large investment projects
- Promoting the functional consolidation of investment operations

HSINCHU SCIENCE-BASED INDUSTRIAL PARK (HSIP): THE 'SILICON VALLEY' OF THE EAST

Among many industrial parks or zones on the island, HSIP was the only one tailored to high-tech industry (Mathews, 1997; Tanzer, 1998). The park was under the jurisdiction of the National Science Council, which had the traditional role of a government funding agency for academic research. Located in Hsinchu County, approximately 80 km to the south of the capital city Taipei, HSIP had easy access to the international airport and harbours, a skilled labour force and abundant technological resources, including two national universities and the government-sponsored ITRI. Since its inception, HSIP has received over US$500 million from the government, earmarked for the acquisition and development of land and construction of housing and factories. The developed land totalled approximately 600 ha (1500 acres) and was divided into distinct industrial, residential and recreation areas.

OVERVIEW OF PARK INDUSTRIES

The number of companies in the HSIP had risen to 245 in 1997. HSIP companies are classified into six categories: integrated circuits (ICs); computers and peripherals; telecommunications, opto-electronics; precision machinery and materials; and biotechnology (see Table 3.6). With an ever-increasing concentration of industrial clusters, HSIP has become the undisputed heartland of Taiwan's high-tech industries. Of the six industries in the park, ICs, computer and peripherals and telecommunications have a competitive edge in the worldwide arena. Although

Table 3.6 Industries in HSIP, 1997

Industry	Firms	Employees	Sales (US$m)	Growth (per cent)[a]
IC	96	37681	6690	21.9
Computers/peripherals	44	17263	4908	11.4
Telecoms	37	4877	945	34.9
Opto-electronics	35	6994	970	52.1
Precision machinery/ materials	18	1295	119	18.4
Biotechnology	15	300	14	56.7
Total	245	68410	13916	20.3

Note: (a) Growth rate is compared with 1996 sales figures.

Source: HSIP administration, 1998.

many industrial analysts would describe Taiwan as the USA's manufacturing backyard, it is nevertheless true that Taiwanese firms in these technical fields, such as Acer and Taiwan Semiconductor Manufacturing Co. (TSMC, see Chapter 5), have become global players. Taiwan rightfully deserves the name of the 'Far East Silicon Island' (see Figure 3.1).

In 1997 combined sales of HSIP firms grew by an average 20.3 per cent, reaching US$13.9 billion. Of the companies in the park, 43 were foreign-owned and 202 domestically owned. Aggregate investment increased by 72 per cent from 1995 to reach US$13.1 billion by the end of 1997, and domestic sources accounted for 88 per cent of park investment capital. By the end of 1997, 84 firms in the park had received permission to increase their capitalisation, the total capital raised amounting to US$4.6 billion; of these firms, 32 were in the IC industry and raised a total of US$3.6 billion in new capital (see Tables 3.7, 3.8).

The activities of the park companies were increasingly geared to internationalisation: 58 companies had offices abroad and many well-known foreign manufacturers signed science and technology cooperation agreements with park companies (see Chapter 2). The California-based Virtual Silicon Technology (VST) and the HSIP-based United Microelectronics Corporation (UMC, see Chapter 6) entered into a long-term joint marketing and technology development agreement, collaborating on developing test chips, establishing timing correlation and verifying Extended Partition (EP) cores. Partnerships like this can only strengthen HSIP's technological leadership position for the foreseeable future.

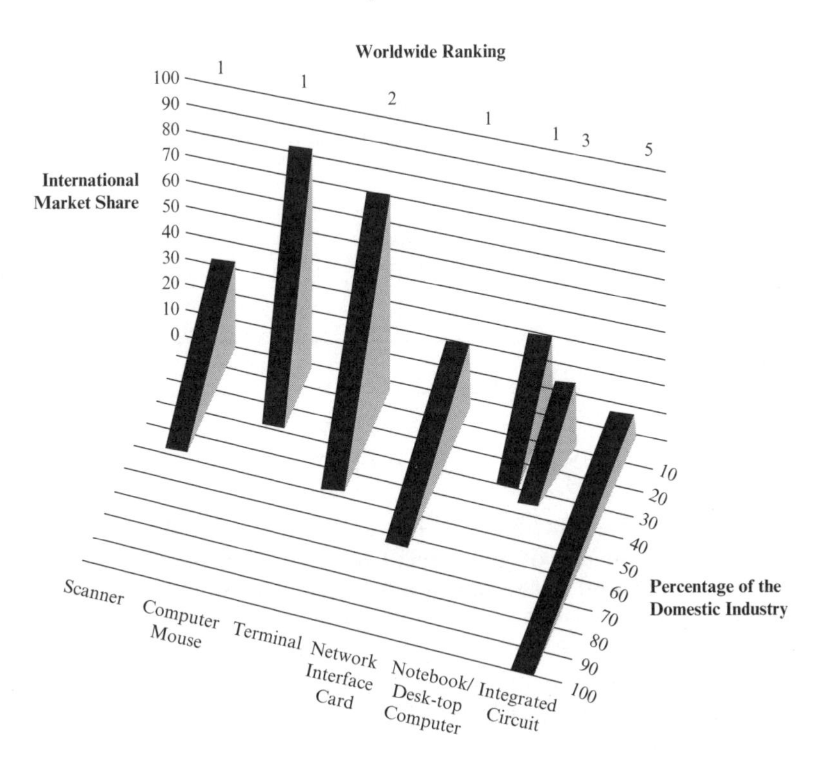

Figure 3.1 Contribution of HSIP industries, domestic and worldwide, 1997

Table 3.7 New investments in HSIP, 1997

Industry	Companies approved	Amount invested (US$m)
IC	22	491
Computers/peripherals	5	46
Telecoms	6	38
Opto-electronics	6	212
Biotechnology	6	118
Total	45	905

Source: HSIP administration, 1998.

According to a survey done by the Science Park administration, the failure rate of park companies was about 10 per cent. Compared with that of industry parks in other countries – as high as 90 per cent – this figure was impressive.

Table 3.8 Firms raising capital in HSIP, 1997

Industry	No. of firms	Capital raised (US$m)
IC	32	3269
Computers/peripherals	14	558
Telecoms	11	118
Opto-electronics	20	285
Biotechnology	1	2
Total	61	4961

Source: HSIP administration, 1998.

HOW ARE CLUSTERS CREATED?

Several factors have contributed to Taiwan's success in building up its high-tech industries: the role of government support, the utilisation of human resources, R&D, and the complete service package provided by the park administration.

THE ROLE OF GOVERNMENT

As we have seen, the role of government can facilitate the building of the infrastructure necessary for high-tech industries. This can be observed from the initial stages of Taiwan's high-tech development. To induce the creation of industrial clusters, the government provided full support, financially and legislatively, for the acquisition of technology from other countries, R&D projects and technology transfer to firms. But it never interfered with the firms' business operations.

Without a doubt, technology was the backbone for building a high-tech industry. The government played an essential role in fostering the flow of high-tech industry through licensing, acquisition and reverse engineering. This can be seen from the establishment of Taiwan's semiconductor industry, to which ITRI and its Electronics Research Service Organisation (ERSO, see Chapter 2) made a remarkable contribution. ERSO was established in 1974 and its primary focus was on semiconductors, display devices, microwave components and electronic packaging. Employing about 600 engineers, ERSO engaged in R&D for the government and then transferred the technologies into the market. Beginning with the acquisition of 7μm CMOS technology from RCA in the mid-1970s, ERSO has been instrumental in building Taiwan's IC industry. In 1980 it spun off

UMC, the first Taiwanese IC and the first high-tech firm to reside in HSIP. In the next two decades, based on the spin-off model, ERSO gave birth to several firms in the park. The Electronics Testing Centre was spun off in 1982 with new IC testing technologies; the Taiwan Semiconductor Manufacturing Co. (TSMC) was the third spin-off when VLSI technology (at 2μm resolution, then trailing the world's best of 1μm) was acquired from Philips; the Taiwan Mask Corp. (TMC), the country's first domestic mask manufacturer, was spun off in 1988; and in 1994, Vanguard International Semiconductor (VIS) was spun off as the first 8-inch wafer fab in Taiwan.

HUMAN RESOURCES

The available human resources (HR) base was another essential factor in maintaining the competitiveness of the HSIP high-tech firms. By the end of 1997 about 68 410 people were working in the park. The proportion of employees with at least a junior or technical college education was 58 per cent, and almost 15 per cent had graduate degrees. This HR endowment gave HSIP a great leverage over its counterparts in Singapore, Malaysia and Korea. Returning expatriates also played an important role in the development of the park, founding 97 companies. The technical and scientific skills and ideas of these 3000 returned expatriates were one of the main driving forces propelling HSIP forward.

R&D

R&D was the cornerstone of HSIP. The companies in the park spent about 5.4 per cent of their revenue on R&D, compared with only around 1 per cent by the manufacturing industry in Taiwan as a whole. With the ambitious undertakings in R&D, the high-tech firms in HSIP were able to maintain their competitive advantage. In the IC industry, for example, the main strength of Taiwan compared to other chip producing regions such as the USA, Japan, Korea and Europe lay in the smooth uptake of R&D output and the proximity of the R&D to the wafer manufacturing units. All the activities were seamlessly integrated in HSIP and the tacit knowledge was disseminated at a rapid speed. This also made Taiwan's semiconductor industry an important part of the global market. To encourage R&D, the Science Park administration presented annual Innovative Product awards and provided grants for innovative high-tech R&D projects and strategic products and components. Since its inception in 1987, this funding system had contributed significantly in raising the global competitiveness of the HSIP firms.

Another key to enhancing R&D capabilities of HSIP firms can be traced to the strategic technology alliances (STA) made with both domestic and foreign companies. According to the Taiwan Semiconductor Industry Association (TSIA), HSIP-based Mosel Vitelic and Siemens established ProMOS Technology together, TSMC produced 16Mb DRAMS for Fujitsu, Macronix had mutual patent licensing with IBM, and Vanguard International cooperated with Sharp, Matsushita and Motorola. STA proved the way for HSIP firms to tap into advanced technologies.

COMPLETE SERVICE PACKAGE

The complete service package provided by the Science Park administration was a fourth factor, but perhaps the most important one, that helped to fertilise the high-tech industry in HSIP. Director Kung Wan said: 'I have visited many science parks in other countries. What I have seen that differs is our complete service, not just a piece of land, not just some tax holidays, incentives, or regulations.'

HSIP services reach into daily life, public recreation centres and education facilities. HSIP established a bilingual school in 1983, which helped to retain high-quality human resources, especially expatriates. Commercial needs were no less skilfully dealt with. HSIP not only provided ready-made buildings, but also helped firms to formulate their operation plans, locate their suppliers and formulate their operation plans, locate their customers, and even recruit their management team (see Table 3.9). In many other countries external consulting firms provide these services. HSIP also provided the so-called 'one-window services' by which a firm could get through all export procedures in one stop. This enabled HSIP firms to make faster delivery to their customers in comparison with Korean and Japanese firms.

But there was no complacency: Director Kung Wang was emphatic: 'We are always learning and we are always improving our services to firms'. HSIP has built sister relationships with its counterparts in other countries, and frequent exchanges with these parks help HSIP keep abreast in improving its services. In July 1995 the Science Park administration completed the installation of a park-wide Asynchronous Transfer Mode (ATM) broadband network, which brought park administrative services, operations management, R&D and networking fully into the information age.

Other services provided by the park network include automated customs clearance (from form submission to visa issuance to warehousing), electronic document transfer, a labour information database, a human resource

Table 3.9 Investment incentives and benefits in HSIP

Tax incentives

Income tax

- A 5-year income tax exemption period for profit-seeking enterprises and science-based industries; science-based industries also have the choice of the first fiscal year for income tax payments and a 5-year period of exemption thereafter.
- A 4-year income tax exemption period on income generated from newly added facilities for carrying our a capital expansion project; or 15 per cent of the cost of the newly added production/servicing equipment may be credited against income tax in the year of expansion (if taxable income from the newly added facilities in that year is less than the applicable credit, the credit may continue to apply in the following 4 years).
- Following the tax exemption period, income tax/surcharges for profit-seeking enterprises shall not exceed 20 per cent.
- When science-based industries recapitalise undistributed retained earnings for procurement/replacement of machinery/equipment/transportation facilities used for production of goods or industrial safety/sanitary standards matters, any shareholders' newly issued name-bearing stock certificates shall be exempt from income tax in the year of recapitalisation (shareholders must pay the income tax liable when any stock certificates are transferred).

Import duties

- No import duties are leviable on raw materials, fuels, supplies, or semi-finished products imported by an HSIP enterprise for its own use; the importer is not required to file for exemption from import duties.

Business taxes

- Business taxes are computed at zero per cent for goods/services exported by an HSIP enterprise.
- Earnings of up to 200 per cent of their paid-in capital may be retained by HSIP industries designated by the ROC government; profits in excess of this are taxed at 10 per cent, the remainder after tax may be retained and is not required to be distributed.

Investors' rights protection

- Foreign and/or overseas Chinese investors enjoy the same rights/privileges as ROC (Republic of China) national investors.
- Foreign and/or overseas Chinese investors may have 100 per cent ownership in an HSIP enterprise; they may also enter into a joint venture with the ROC government or local enterprises.
- Foreign and/or overseas Chinese investors may apply to have capital gains/interest and profits from their investments remitted overseas.
- The Taiwan government guarantees that an HSIP enterprise will not be expropriated for 20 years.

Table 3.9 (continued)

- A foreign investor may apply for a one-time outward remittance of capital investment one year after completion of an investment project.
- Intellectual property rights and ownership rights are protected by law.
- With approval from the Science Park administration, science-based industries may engage in import/export activities related to their principal business.
- Science-based industries may invest in other companies in the ROC and abroad on completion of their HSIP investment.

Know-how/patent rights
An investor may contribute technical know-how or a patent right as an equity investment which, singly or in combination, cannot exceed 25 per cent of total capital; non-patented technical know-how must be matched by an equal or higher amount in tangible assets or cash and may not be transferred to any other party for a period of two years commencing on the date of completion of the investment project.

Loans
Science-based industries may apply to the Bank of Communication for a low-interest loan for construction of factory buildings/purchase of equipment; the interest rate will be approximately 2 per cent lower than the usual primary rate for bank loans; the amount of the loan must not exceed 30 per cent of the capital required for the buildings/equipment or 65 per cent of the cost of the investment project. The maximum loan repayment period is 10 years, including a grace period of 1–3 years.

R&D facilitation
The Science Park administration may provide grants for technological R&D activities undertaken by registered science-based industries in HSIP; as part of the application each industry must submit a comprehensive R&D activities plan covering the next five years. If approved, the industry may receive up to NT$5m for each R&D project: the amount of the grant may not, however, exceed 50 per cent of the project's total cost.

and vehicle information system, the National Science Council's technology database, an electronic bulletin board, World Wide Web access, a multi-media guide to the park, multi-point video conferencing, and distance learning applications.

TAINAN SCIENCE-BASED INDUSTRIAL PARK (TSIP)

In an attempt to upgrade high-tech industry in Southern Taiwan utilising HSIP's experience, the Executive Yuan decided in 1995 to set up TSIP in

Tainan County. The area covered about 638 ha, and the target industries were semiconductors, microelectronics precision machinery and agricultural biotechnology. TSIP is located near the Sun Yat-sen Freeway and the main railway line, and is about an hour's drive away from Kaohsiung International Airport and Kaohsiung Harbour. Close to the park are many academic and research institutions including the National Cheng Kung University, the National Sun Yat-sen University, the Asian Vegetable Research and Development Centre and the National Yulin Institute of Technology.

TSIP began to accept applications from companies in March 1996. In the first phase, TSIP was to be developed into three specialised zones to host microelectronics and precision machinery, semiconductors and agricultural biotechnology industries. In addition to supporting investment in introducing new high-tech industries, TSIP planned to establish communal facilities equipped with advanced instruments. Efforts would also be made to promote cooperative research projects between academe and industry. Awards and incentive programmes were put in place to help establish an environment conducive to the growth of high-tech industries. To the end of 1997, a total of 44 companies had been approved, including 27 companies from HSIP. Total investment was expected to reach US$60 billion in 2010. Four companies including TSMC already had commercial construction.

THE CHALLENGE AHEAD

The Impact of 'Asian Flu'

HSIP has been regarded as a centre of Taiwan's national high-tech industry, and many companies continue to apply for a place in this high-tech haven. The semiconductor industry of the park has established vertical alliances which makes HSIP the world's fourth largest production centre for semiconductors. The computer and peripheral, opto-electronics and telecommunication industries are also growing at a steady pace: 'In the face of Asian economic crisis, Hsinchu industries played a critical role in keeping Taiwan away from the flu', according to Director Kung Wang. Taiwan's economy grew 6.00 per cent in the first quarter of 1998, faster than many economists had predicted and in line with the government's prediction of 6.18 per cent. Taiwan had thus ridden out much of the turmoil that had shaken Asia for the previous ten months. Driving the growth was a larger than expected increase in private investment, which helped to offset the drop in exports. The Science Park administration was undertaking a fourth expansion plan for HSIP, involving a total of 512 ha and it was expected that the number of companies and employees in the park would double by 2006. Output value and expenditures for R&D were expected to rise fivefold to reach US$52 billion and US$2.3 billion respectively. However there exist challenges, many of which originate from the change of the global economic environment and the resource constraints of the island. Taiwan, until early 1998 an island of tranquility in the Asian economic typhoon, was about to put up warning flags. Virtually all of Taiwan's fabs cut their capital expenditure budgets in the face of the falling demand for ICs; as expansion plans were scaled back, current equipment utilisation rates in Taiwan's existing factories ran anywhere from a poor 70 per cent to a feeble 50 per cent. It was estimated that Taiwan's total semiconductor capital spending would drop 5–10 per cent from the 1997 US$5 billion level. That was a reversal of earlier forecasts, which had projected Taiwanese companies increasing their spending by 10–15 per cent in 1998, to as much as US$6 billion. Now it appeared that total Taiwanese capital spending in 1998 could be as low as US$3.9 billion.

Internal and External Challenges

A number of other challenges had also begun to arise following Taiwan's period of rapid progress. Labour costs had increased, real estate prices had soared, industrial land was hard to find, labour-intensive industries were moving offshore and environmental protests had become a common occurrence. Taiwan was now also faced with a new external situation with the formation of regional economic blocs around the world, intense international competition and the more liberal economic and trade system brought by the conclusion of the Uruguay Round of GATT talks and the creation of the World Trade Organisation (WTO).

For HSIP, the first challenge that the Science Park administration had to face was the difficulty in further land acquisition. Another was the shortage of experienced engineers. A manager of an IC firm in HSIP said that 'was probably one of the most acute problems that Taiwan's semiconductor industry has to solve in the next decade'. Taiwan can produce good quality engineers at the entry level, but more experienced engineers are difficult to find. In the past the industry relied heavily on overseas Chinese and on people trained by R&D organisations such as ITRI. But growing companies had gradually depleted the pool of resources, and coming up with a new source of experienced engineers was proving a real problem.

Challenges also came from the increasing numbers of science and technology parks in surrounding countries including Malaysia, Singapore and mainland China. There were only 22 such parks on the mainland in 1994, but the number had doubled by the end of 1997. The parks on the mainland caused concern because they not only had more attractive terms and incentives but also had a strong human resources supply and huge domestic markets. All these were also threats to the development of TSIP. Although the Tainan area was the main base for traditional industries, its high-tech foundation was analogous to that of HSIP in 1980. The inadequate location and infrastructure seriously affect Taiwan's attraction for high-quality human resources.

Director Wang provided a summary of the response to these challenges.

> Taiwan had no other choice but to move forward. Technology is the fuel that powers economic development; advanced technology is the mark of a nation's strength. During the past sixteen years, the Hsinchu Science-based Industrial Park has played a leading role in reshaping Taiwan's industrial base and bringing about the development of high-value-added products. What we are thinking about now is the way to sustain our competitive advantages.

CONCLUSION

In 2002 Taiwan was standing at a crossroads. Although there were opportunities for future development, threats had emerged forcing both the government and industry leaders to have a serious strategic rethink. HSIP had run out of space, and environmental factors in Taiwan created little optimism. The institutional scene had changed; environmental activism was in full flower and building a replica of HSIP in Taiwan's rural south would not be as easy as before. Labour was no longer cheap compared with Taiwan's neighbours and experienced engineers expressed their reluctance to relocate southwards. Other major liabilities included a small domestic market and few natural resources, both of which seriously challenged the OEM-based model of Taiwan's high-tech industries. Since 1990 some 80 000 businesses moved offshore, about half to China and half to Southeast Asia, and in 2002 more than half of Taiwan's electronics production was off the island: people were starting to worry about Taiwan's 'hollowing out' and for a government perennially concerned about national security, watching a growing proportion of the economy move out of its reach was frightening. But the number one threat to Taiwan was the political tension between the island and mainland China. The market crash of 1996 had to a certain extent faded from the memory of business people, but a new round of military confrontation between the two sides of the Strait looked imminent.

NOTE

For teaching and reference purposes, a companion case *Science Parks in Taiwan* (9A98M016) can be obtained from Ivey Publishing, Richard Ivey School of Business, University of Western Ontario, Canada. The authors are particularly grateful to the Jean and Richard Ivey Fund which provided support for the research fieldwork.

REFERENCES

Mathews, J.A. (1997), 'A Silicon Valley of the East: Creating Taiwan's Semiconductor Industry', *California Management Review*, **39**(4), 26–54.
Tanzer, A. (1998), 'Silicon Valley East', *Forbes*, June 1.

4. Macronix International Co. Ltd (MXIC)

Chin-kang Jen

TAIWAN IN THE WORLD SEMICONDUCTOR INDUSTRY

In 1999 the total capacity of Taiwan's semiconductor industry constituted 10.9 per cent of world market, only marginally less than that of the USA and Japan, and therefore rated third in the worldwide market (see Table 4.1).

Among the 21 manufacturing companies, most were professional wafer foundry companies, and only a few were integrated device manufacture (IDM) companies that had their own fab and focused on design and development, manufacturing and sale for their own brands; a totally different situation to that of the foreign semiconductor industry, which consisted mainly of IDM companies. This was not only because of the prevalent

Table 4.1 Taiwan's semiconductor industry, 1999

	Per cent of worldwide output value	World position	Value (US$)
Total output value	4.7	4[a]	
IC design	19.6	2[b]	2.3 b
IC manufacturing	6.8[c]	4[a]	8.2 b
(Wafer foundry)			
(DRAM accounting)			
IC package	29.3	1	2.0 b
IC testing	28.0	1	57.2 m

Notes:
(a) After the USA, Japan and Korea.
(b) After the USA.
(c) Wafer foundry and DRAM account for 53 and 33 per cent of total output value, respectively.
(d) Factory number: Design companies: 127, Mask ROM: 5, Manufacturing: 21, Package: 42, Testing: 33.

habit in Taiwan's industry of starting a new enterprise to exploit a new technical development, but also because the enterprises were good at vertical specialty division. It was also related to the 'aggregation effect' of HSIP (Chapter 3) and to the flourishing cooperation network existing between enterprises.

However even in an industry environment with prevalent foundry services, one enterprise still insists on its own approach and never provides foundry services. As an IDM, they develop their own brands; there is no DRAM, only non-volatile memory products. This alternative approach has been developed by Macronix International Co. Ltd (MXIC), which in 1997 had 37 per cent of market share and first position in the world in MASK ROM, and twelfth position in the world in market share of Flash memory.

MXIC'S ENTREPRENEUR

Miin Wu was on the cover of *Forbes* in August 1999. After gaining the degrees of Bachelor of Electrical Engineering at National Cheng Kung University and Master of Materials Science at Stanford University, Wu worked for eight years for several companies such as Siliconix, Rockwell, Intel and VLSI: many entrepreneurs in 'Silicon Valley' were cultivated from Stanford University. At Stanford University Wu's approach was very different from general entrepreneurs with an engineering or R&D background, as he was determined to gain as thorough a knowledge as possible of plant building, sales and the challenges of the external environment for companies.

CREATION OF THE MXIC FACTORY

The development of Taiwan's IC industry began with the establishment of the Kaohsiung Electronics Company in 1966 but for many years it remained at the assembly stage. The real start-up of manufacture began when the Industrial Technology Research Institute (ITRI, a forerunner of the Electronics Research & Service Organization, ERSO, see Chapter 2) was founded in 1974. ITRI was authorised by the Ministry of Economic Affairs (MOEA) to carry out an IC model factory establishment project, completed in 1977. Owing to the lack of hands-on experience, ERSO continuously sent staff to the USA to learn new technology. In his job of implementing technology transfer, Wu became aware that there was a chance for East Asia to exploit the coming technological revolution. In 1987 and 1988 Taiwan's stock market rocketed and capital seemed abundant. The productivity of Taiwan's semiconductor industry, and its capital and human resources, was close to maturity. Wu took Ding-Hua Hu from H & Q Asian Pacific as his partner, who had taken part in the IC model factory establishment project while working at ERSO. The focus in semiconductors in the mid-1980s was only on the manufacturing process, and Hu approved Wu's idea to set up a factory to work from process development to production and create proprietary products, an approach that could not only strengthen Taiwan's IC manufacturing capability, but also its IC design capability.

In December 1989 MXIC was established with capital of NT$0.8 billion. The stockholders included individuals and corporations such as Delta Electronics Inc., Mercuries & Associates Ltd, Chiao Tung Bank, Picvue Electronics Ltd, as well as H & Q Asian Pacific.

STRATEGY AND NICHE

When MXIC was first established Wu was already determined that the company should go the way of system integration. However since MXIC was a new company that had yet to build up the R&D of logic IC, Wu decided to make memory chips MXIC's main product. In the longer term, if MXIC was intended for system integration, what kinds of systems should be provided? What were the internal needs within the system? Wu thought that the final goal for system integration should be to provide products that were slim, short and small, so the greatest potential would lie in hand-held products needing low power. With volatile memory, such as DRAM and Static Random Access Memory (SRAM), for example, if the power is turned off the data in memory will vanish, so the power must be kept turned on to retain the data. With non-volatile memory (NVM), such as Mask Read Only Memory (MASK ROM), Erasable Programmable Read Only Memory (EPROM) and Flash, data can be saved even if the power is turned off; when the power is turned on, the data can be read from the NVM again. The data in MASK ROM is saved during the process of making the wafer; in EPROM and Flash, the data is written after the IC manufacture is completed. For clients MASK ROM thus has the least flexibility but the cheapest price. In order to provide a superior system integration service MXIC needed to be equipped with capabilities of NVM design and manufacture. Because Flash was just starting in the market, MXIC chose the simplest MASK ROM as the entry item, then EPROM and then Flash. Since MASK ROM already had a certain market at that time, operation risks could be minimised if MXIC could design the product on its own and sell the product by itself.

Wu understood that building the factory, establishing the technical development process and cultivating clients would all take time, and only when the sequence of these three events had been completed would there be a real enterprise. This might take five or six years, in which time the company might have disappeared owing to lack of profit. Wu determined to carry out the three events simultaneously: establish relationships with clients and start product development at the same time, and then build the factory. To find the clients MXIC needed to know which companies would need its solutions. After locating the clients, the next step was to build a relationship with them and provide solutions for them. The staff responsible for product development would then have to start product development based on clients' needs. When the production line was created, the developed product would have to be moved to the production line. The biggest difficulty was that MXIC had no production line and no products, so how could it find its clients?

The excellent quality of Japan's technology products in the mid-1980s meant that they were prevalent throughout the USA and had a great impact upon the USA's own factories, and the Semiconductor Industry Association (SIA) lobbied the US government to impose pressure on Japan to make 20 per cent of the Japanese market open to foreign companies. When MXIC was first established, Wu had also invested in an alliance with another US company so he could use its name and trademark to pretend that MXIC was an American company. Owing to the poor quality of products, most American companies were withdrawing from the MASK ROM market. At that time, the biggest client for MASK ROM was Nintendo from Japan; to prevent copyright problems, all the cassettes within game machines produced by Nintendo were using MASK ROM, and for Nintendo MXIC was just a new unknown company. If 20 per cent of products could have MXIC's trademark, it could meet the US government's requirements.

The problem about clients had been solved, but what about the product? Wu found Samsung Electronics in Korea; although Samsung Electronics had a MASK ROM product, it had no chance to enter the Japanese market because Japanese companies preferred to adopt their national products. Using MXIC's product, Wu could enter the Japanese market and cooperate with Samsung Electronics to sell Samsung Electronics' product with MXIC's trademark. Of course, Wu was aware that Samsung Electronics' intention was to start the business with Nintendo by using MXIC, then tell Nintendo that all the products came from Samsung Electronics and do business directly without MXIC. As soon as he had sold the first products to Nintendo, Wu sent an R&D team to Chartered in Singapore, finished the product development, and gradually started mass production. This

Table 4.2 MXIC in the worldwide MASK ROM market

- **Good product architecture**. MXIC developed its own ROM unit architecture in flat cell format. This format has the advantages of small surface and short production period. MXIC can use this unit architecture in new product development and therefore design and production experiences are accumulated.
- **Making MASK ROM the core competence**. The major difference between MXIC and SHARP, NEC and Samsung is that MXIC takes MASK ROM as its most important core competence, so all the most advanced process technology will be used in the mass production of MASK ROM. Other major competitors use their most advanced technology on their own core products instead of MASK ROM since that is not their core competence.
- **Good production management**. Twelve masks are needed while producing MASK ROM, and within them ten masks are needed for forming ROM Code. In the low season, MXIC allocates some part of capacity for producing semi-finished products with ten masks, and the other capacity will be used for producing standard products such as EPROM and Flash. In the high season, all the capacity will be focused on MASK ROM and completing the two masks for the stored semi-finished products.
- **Compressed delivery date**. The so-called turn-around-time (TAT) is the period between the client giving its ROM Code to the manufacturer and it receiving the completed product. In the early days, MXIC's TAT was 40–50 days; now it takes 12–14 days. Besides semi-finished products, MXIC has simplified its production flow, facilitating improved production flow of collaborative companies, consigned mask machine and invested in a packaging/testing factory, all with the aim of consolidating the supply chain and promoting production efficiency.

Source: *DigiTimes* (2000).

was a preventive strategy in case Samsung Electronics later chose to terminate the contract with MXIC. With the completion of its own factory, Wu moved the production line back to MXIC. MXIC took first position in the worldwide MASK ROM market for four main reasons (see Table 4.2).

SYSTEM INTEGRATION AND CLIENT RELATIONSHIPS

System integration has two levels of meaning. The first involves the system integration of technology – in other words, putting all ICs with different functions together on the same chip, for example MXIC's

System-on-a-Chip (SoC). The second involves the system integration of the entire service for the client – MXIC not only provides production but also provides the total client solution for marketing, branding and design. MXIC defines itself as an Integrated Solution Provider (ISP) instead of a pure IDM company.

If the solution MXIC provides to a client is the most advanced product in the field, then naturally MXIC creates great added-value for the client. However if MXIC wants to provide the most advanced product, the client has also to be the most advanced company. MXIC has therefore to find the most advanced client among those using NVM.

Wu regards return on investment (ROI) as a very important indicator of MXIC's management competence. Since MXIC aims at providing a system integration service, it will provide a custom-made solution for the client's need: the concept of a *client-driven service*. In order to provide a total solution for a client's specific need, Wu believes that the requirement is to find the biggest client, with the biggest share of the world market. When MXIC completes the product that the client can use, the client will naturally have a great demand for MXIC's supply, and MXIC gets a significant return on demand. How to choose a good client is thus very important to MXIC.

For MXIC a client with large market share will bring big deals, and so MXIC has to assure the client of a continuing supply of product. If MXIC adopts an out-sourcing approach, the quantity to be supplied will be controlled by wafer foundry factories: MXIC's commitment will be incomplete. To create integration for an enterprise it is necessary for MXIC to have production control. To provide a system integration service, MXIC has to be equipped with capacity for design, manufacture, testing and service.

When MXIC began contacts with clients, they were very picky and their orders were only of small quantity. So MXIC took a Key Account approach to find important clients, establish long-term relationships with them, quickly grasp their core products, and then be a major supplier. As the relationship with clients got stronger, the clients would be willing to tell MXIC their product plans for two or three years into the future and would want MXIC to design products according to their own specifications. MXIC would thus get information about clients' product plans earlier than other competitors.

Why were clients willing to give such sensitive information to MXIC? It is because the clients can design their own systems, but they have no idea of which ICs can be used to achieve the desired effects. During the system design phase, for example, the clients will consider how fast the speed of graphics process and how fast the speed of the CPU should be.

Clients are usually good at these kinds of decisions, but to design the IC is MXIC's specialty and clients have to depend on MXIC for this. For systems designed by clients MXIC will also provide a professional opinion on IC design and assess if it is possible for current IC design technology to achieve the clients' requirements: MXIC's team becomes a member of the clients' R&D teams. When Nintendo is designing an N64 game machine, it will first tell MXIC the N64's functions; MXIC will then design a Rambus clock chip using advanced technology. At the same time, a new ROM will be designed based on MXIC's specialty; Nintendo's cost will be vastly reduced and MXIC will be the only vendor supplying this type of IC. In recent years Nintendo has been the biggest user of MASK ROM in the world, so MXIC gets the greatest worldwide market share.

Table 4.3 shows that MXIC's first five clients were all world-famous companies. MXIC's clients also include Motorola, SONY, NEC, Infineon, Toshiba and Palm – proof if proof were needed that MXIC's technology and service has achieved world-level quality.

FOCUSING ON SELF-DEVELOPED TECHNOLOGY

Because MXIC takes the self-developed technology approach and has its own products, it can accumulate the technology on integrative products and the intellectual property (IP) will also be accumulated continuously. As soon as MXIC was established, it transferred IC design technology to NKK (a Japanese train company) and this earned US$16 million for MXIC in the first three years of the cooperation. Table 4.4 shows that though technology royalties did not create a significant proportion of overall profit, it still illustrates the importance of MXIC's self-developed technology strategy.

Table 4.5 shows that MXIC spent more than NT$1 billion on R&D and this expense consumed about 10 per cent of net profit. In addition to the budget for R&D, MXIC does not take all wafers for production; it always leaves a certain ratio of wafers for the R&D department and rewards staff who get IP patents.

If clients want the integrated system to be implanted into SoC, the technology will have a high degree of complexity. It is very difficult for general vendors to be equipped with the required IP, so it is necessary to cooperate with many vendors having different IP. Table 4.6 shows MXIC's cooperation partners, most of them famous in the field, indicating that MXIC's IP already has a significant level of competence.

Table 4.3 MXIC's first five sales clients, 1996–2000, per cent of market share

1996	%	1997	%	1998	%	1999	%	2000	%
1 Nintendo	31.50	Nintendo	48.93	Nintendo	53.51	Nintendo	49.57	Nintendo	34.29
2 TSMC	4.68	Inventec	4.24	HP	3.19	HP	3.79	HP	9.17
3 Inventec	3.52	MEGACHIPS	2.16	Inventec	2.28	Panasonic	2.89	Mitsubishi	8.33
4 DAIWA SANKO	2.77	ECS	1.72	GI	2.07	Philips	2.37	Philips	1.30
5 MCC	2.63	TSMC	1.63	Philips	1.64	Inventec	1.81	Samsung	1.29

Note: ECS – Elitegroup Computer System.

Table 4.4 IP royalties: percentage of product, 1996–2000

Products	1996	1997	1998	1999	2000
MASK ROM	52.55	56.76	65.00	63.85	59.82
EPROM	25.01	12.44	10.61	7.75	3.79
Logic	9.03	15.08	10.60	11.89	5.91
Flash	8.48	14.30	8.67	12.47	19.93
Foundry	4.49	1.38	5.10	2.98	9.55
Technical Royalty	0.44	0.04	0.02	1.05	1.00

Table 4.5 R&D expense (NT$ million) and percentage of net income, 1996–2000

1996	%	1997	%	1998	%	1999	%	2000	%
NA	NA	1427	13.87	1977	16.05	1945	11.71	3139	9.74

LISTING OF SECURITIES

MXIC started its operation using capital of about NT$0.8 billion. The need for huge capital to build a wafer foundry factory naturally imposed great pressure on MXIC's finances. How to successfully raise enough money from the capital market became an important issue for ongoing operations. At that time, all the listed stocks in Taiwan belonged to the first or second categories; in order to protect investors' rights, the regulations and policies Securities and Futures Commission Ministry of Finance, ROC (SFC) was very rigid about the listing of securities. This was a great disadvantage to a hi-tech company that has to invest large amounts of money for equipment purchase. Owing to depreciation, it's not easy to attain a profit and loss (P&L) balance in the first few years, not to mention any surplus. In Wu's estimation based on the rules at the time, it would take MXIC at least seven or eight years to get a listing of securities. This could not satisfy MXIC's urgent needs.

Although the rules about the third category – technology stocks – had been passed for seven or eight years, the government never approved any company to be listed as a third-category stock because it feared the possibility that if a company was listed before it made any profit, it would affect investors' confidence if the company later closed down. Wu explained to the MOEA and Ministry of Finance officials that hi-tech companies had to be listed as soon as possible in order to get enough capital to flourish and

Table 4.6 MXIC's main collaborative partners, 1997

Partners	Collaborative issues	Collaborative items
NKK (Japan)	Technology transfer/product development	Flat ROM, Flash
Sanyo (Japan)	Authorisation	MASK-SET
Megachips (Japan)	Permission for production/sale	AFIC
Matsushita (Japan)	Long-term strategic cooperation/production (strategic alliance)	DRAM
Mitsubishi (Japan)	Long-term strategic cooperation, R&D, production/sale (strategic alliance)	SRAM, Flash
ACE Logic (USA)	Product development	PC Based GUI/Video
RAMBUS (USA)	Authorisation	Rambus Channel
MIPS (USA)	Technology transfer	MIPS R3000, CPU
VLSI (USA)	Collaborative operation/production (strategic alliance)	Embedded Flash
Philips (Netherlands)	R&D, production/sale (strategic alliance)	Embedded Flash
L & H (Belgium)	Product development	Speech Coding
CCL (Taiwan)	Product development/technology transfer	ISDN S/T Chip, MEPG-2 Audio Decoder
ERSO (Taiwan)	Technology transfer	DSP-IC

mature. The government finally decided to make third-category stocks open to the public, and in March 1995, approved by the Industrial Development Bureau, MXIC became the first company listed as a third-category – technology stock – in Taiwan.

In addition to getting capital from Taiwan, MXIC also tried to get money from abroad. Wu considered the NASDAQ to be the best and most transparent capital market in the world, so if MXIC could be listed on the NASDAQ, it could not only get foreign capital but it would also indicate that MXIC was a very transparent company with certain level of technology. This would benefit MXIC in its interaction with the best companies in the world. In May 1996 MXIC successfully became the first company in Taiwan listed on the NASDAQ, by way of American deposit receipt (ADR).

CONCLUSION

In Taiwan's semiconductor industry, MXIC is an extraordinary company, with behaviour – including product strategy, investment in R&D, and listing of securities – that is very different from the more general semiconductor companies in Taiwan. MXIC thinks that there are four requirements for being a good ISP – an abundant IP library, a capacity of integrated design on system integration chip, a wafer manufacture factory powered by advanced process technology and first-rate international collaborative clients. MXIC possesses all four requirements.

REFERENCE

DigiTimes (2000), 'Operating system framework: MXIC linked up key account strategy with supply chain to implement knowledge management by complete informational flow system', www.digitimes.com.tw, 11 November 2000.

5. The model of Taiwan's high-tech industry: TSMC

Chia-wu Lin

The new millennium began poorly in Taiwan, with a decreasing economic growth rate and increasing unemployment. The hi-tech electronics industry in Taiwan, and TSMC which leads it, is also suffering from shrinking turnover and unceasingly gloomy financial predictions in conditions of worldwide depression and slowing demand for the computer peripherals and semiconductor industries. Though facing a deteriorating external environment, TSMC still went through with a merger with Acer to create TSMC-Acer and Worldwide Semiconductor Manufacturing Corp. (WSMC), actively established in the Tainan Science-based Industrial Park (TSIP, see Chapter 3). What is the real profile of TSMC? What kind of competition and environment do they face? How do they treat their competition and environment, and how do they respond to them?

PROFILE OF TSMC

> Integrity; Maintaining a Consistent Focus on our Core Business: IC Foundry; Globalisation; Long-term Vision and Strategies; Treating Customers as Partners; Building Quality into all Aspects of our Business; Unceasing Innovation; Fostering a Dynamic and Fun Work Environment; Keeping Communication Channels Open; Caring for Employees and Shareholders, and Being a Good Corporate Citizen. ('Doing Business with TSMC', www.tsmc.com/english/tsmcinfo/c0104.htm)

This is TSMC's operating philosophy, very unlike the brief and sloganised philosophy of most other companies in Taiwan, and reflecting the niche position of the whole company's strategy, market, product and management. TSMC, as described by its manager Mr Kuo-Ding Lee, is a company providing a 'wafer foundry service'. It supplies a professional IC foundry service to semiconductor factories all over the world. TSMC was established in 1987 and funded mutually by Philips in the

Netherlands and the Taiwan government. It is located in the Hsin-Chu Science-based Industrial Park (HSIP, see Chapter 3), Taiwan's 'Silicon Valley'.

PROFESSIONAL IC WAFER FOUNDRY SERVICE

TSMC insists that they do not design or manufacture their own brand of IC products but purely supply a professional 'IC wafer foundry service'. By the end of 2000, TSMC had set up two 6-inch wafer fabs (fabs 1 and 2) and six 8-inch wafer fabs (fabs 3–8) (see Table 5.1).

TSMC began to build 12-inch wafer fabs from 1999. The first 12-inch wafer fab, the sixth wafer fab to be located in TSIP, has begun to receive orders for mass production; the second, located in HSIP, was also expected to provide mass production in the fourth quarter of 2000 and there is a third 12-inch wafer fab at the preparation stage in TSIP. TSMC also coinvests with customers to establish similar professional IC foundry service companies – WaferTech in Washington State and Systems on Silicon Manufacturing Company (SSMC) (a joint venture with Philips in Netherlands and one private company in Singapore).

MASS PRODUCTION PROCESS TECHNIQUES

TSMC provides IC mass production process techniques, including the CMOS Logic process, the analogical/digital mixed-signal process, the single chip or embedded memory production process, the Bi Complementary Metal Oxide Semiconductor (BiCMOS) process and the copper process. Its R&D plans were impressive (see Table 5.2).

Table 5.1 TSMC: 'vital statistics', 2001

Item	No.	US$ million	Pieces (million)
Employees	14000		
Operating income		166228	
Profit		65106	
Manufacturing capacity:			
8-inch wafer (1999)			1800
8-inch wafer (2000)			3400
8-inch wafer (estimated 2001)			4500

Table 5.2 TSMC's R&D funding, 1998–2000

Year	$US billion
1998	1.9
1999	2.3
2000	5.1

With these huge investments, in fact, TSMC is recognised as the very first professional IC manufacturing service company with IDM authorisation. TSMC wants to be the leading wafer foundry service and to keep getting ahead of ITRS (the International Technology Roadmap for Semiconductors). TSMC also focuses on obtaining intellectual property (IP) rights and endeavours to get patents in the USA and Taiwan. In 2000 TSMC acquired 523 patents in semiconductor process technology in the USA and 524 patents in Taiwan. This focus on obtaining patents increases the company's dominance in negotiating with other companies about cross-licensing of IP rights. TSMC first announced 0.13μm mixed-signal CMOS process technology in March 2001, and successfully produced 12-inch wafer with this technology in April. TSMC then applied this 0.13μm process technology to core logic, high-speed, low-power microprocessors. Based on TSMC's internal evaluation, they are already one year ahead of ITRS. In the fourth quarter of 2000, the products in the 0.25, 0.18, 0.15 and 0.13μm process made up more than half of TSMC's profit. TSMC is also one of the companies participating in the International Sematech and 12-inch wafer project (13001μm).

FEATURES OF AN IC FOUNDRY SERVICE

Taiwan has been a member of the semiconductor manufacturing industry since 1966. However it was started only from a package process, the second half of the value chain in the semiconductor industry. After TSMC was established, the trend of decomposing the chain in Taiwan's semiconductor industry began; there were even reports that TSMC and UMC were arguing about which company was the first pioneer to propose the vertical division system. The unique model of vertical division of work in Taiwan's semiconductor industry rapidly adapted every phase of Taiwan's professional foundry enterprises to the competitive environment in a swift transition and rapid capital expansion.

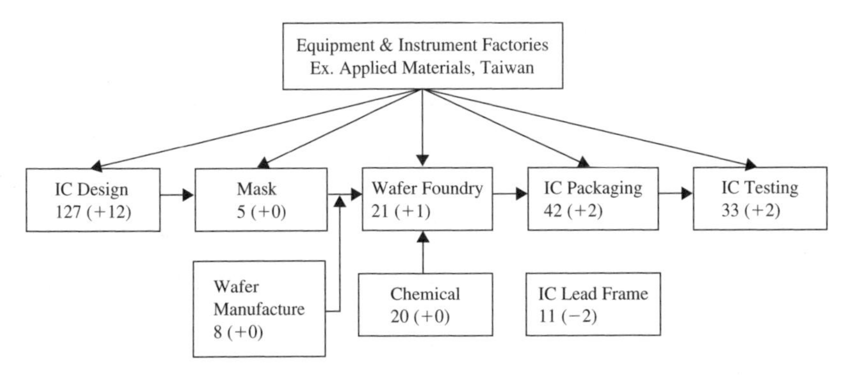

Notes:
The figure before parenthesis means number of companies. + stands for increase in number in 1999, and – indicates decrease of number caused by merger.
There were 16 professional IC Testing companies.

Source: ITIS Program (Apr. 2000).

Figure 5.1 Division of work in Taiwan's semiconductor industry

DIVISION OF WORK

What is the real structure of the professional division of work in the semiconductor industry? The semiconductor industry can be divided into 'IC design', 'IC manufacture' (including wafer foundry), 'IC packaging' and 'IC testing' (see Figure 5.1). Because many IC design-dedicated companies do not have their own wafer fab, they have to outsource their manufactures and other services to alliance partners with professional specialties. In the past few years, the system and IDM companies have focused on their core business and released orders only gradually. In this kind of industry model, the professional division of work in Taiwan has a great impact on output value. Table 5.3 shows that whether measured by product or value chain Taiwan's semi-conductor industry has achieved outstanding performance. In 1999 the output value of MASK ROM achieved first place in the world, and fourth in SRAM, DRAM and general IC products. With regard to the professional foundry manufacture and package processes, Taiwan was first in the world (see Table 5.3).

THE WAFER FOUNDRY

What are the key featurees of TSMC's wafer foundry? The wafer foundry industry is essentially one of high-density capital. According to TSMC's

Table 5.3 The market shares of Taiwan's IC industry in the world (supply), 1999

	Output value (US$ Million)	Market share in world %	Leading position in world	Leading countries
All IC	6146	4.7	4	USA, JPN, KOR
DRAM	3002	14.5	4	KOR, JPN, USA
SRAM	286	6.1	4	JPN, KOR, USA
Mask ROM	359	43.0	1	TWN
IC Design	2295	19.6	2	USA
IC Manufacture	8194	6.8	4	USA, JPN, KOR
Wafer Foundry	4343	64.6	1	TWN
IC Packaging	2038	29.0	1	TWN
IC Testing	572	28.0	–	–
Manufac. Capacity	–	10.9	3	JPN, USA

Source: ITIS program (April 2000).

internal assessment, US$2.00 of assets will on average generate business of about US$1.00. Large PC factories, however, generally need less than US$0.10 to get business of US$1.00. IC design companies without wafer fabs need average assets of about US$0.20 to generate business of US$1.00. In such conditions, and if the debt rate is also high, a professional wafer foundry factory will have to maintain a gross profit of at least 40 per cent to have surplus capital to build new fab facilities, to do R&D on the process technology, to maintain dominance in technology, capability, and efficiency and simply to keep on growing.

BUILDING NEW FABS

The speed of building new fabs is also a key factor in success. To build an advanced 12-inch wafer fab will cost about US$2–3 billion, triple the cost of an 8-inch fab. So why not build 8-inch fabs? Each 12-inch wafer can be split into more chips than can an 8-inch wafer (more than 2.25 times) with less cost (a reduction of about 30 per cent). Investment in 12-inch wafer is mainly motivated by a 'war of cost': for the charting chip companies that produce large quantities of easily standardised large size CPU chips, if they don't have 12-inch wafer fabs, the cost of each memory chip will be 30 per cent more expensive than for companies that do. In the wafer foundry industry, 'efficiency' and 'capacity' are the main 'capital' TSMC used to compete with large worldwide semiconductor factories. In order to retain the position of the largest wafer foundry factory in the world, the general manager of TSMC, Mr Tseng, knows that they will have to 'build one wafer fabrication facility each year!' In addition, 'rate of good product' and 'management effectiveness' are important keys to dominance in competition. Particularly during a depression, saving cost through the rate of good product, efficiency and sound management is essential.

HIGH PROFIT AND HIGH RISK

The semiconductor industry is a business of both high profit and high risk. During prosperity it is common to earn heavily; in a depression or setback in competition, a US$5 billion loss is also common – 'big rise, or big fall' is a normal in this industry. As Mr Yang (the general manager of Winbond Electronics Corporation) says: 'It's not easy to get into [the] semiconductor industry, but it's more difficult to get out of it.' Once in, the vast investment each year will make it impossible for the company to get out. If wafer foundry companies grasp the moment, they can make big money; better still is to be able to preserve their strength while facing low water. Even TSMC cannot be exempted from the current wave of worldwide recession. The diminishing demand for computers and telecoms all over the world, the weakened market caused by America's slow economic growth, and the failure to win more orders from large IDM companies all caused a 17 per cent decline in turnover from the second quarter of 2000 to the second quarter of 2001 and forced TSMC to cut its financial forecast for 2001.

COMPETITION TRENDS IN THE MARKET

Based on TSMC stockholder reports for the first and second quarters of 2001, the company's core business was in trouble (see Table 5.4).

In addition to the internal issues of market depression, surplus of production capacity and vastly reduced orders from customers, TSMC had also to face competition from the large IDM factories. IBM, for example, had the ambition to sew up the market in wafer foundry through its superior capacity and technology. Although Tseng has stated in public that 'IBM targets its customers with high-level products and this will not affect immediately the orders for Taiwan's companies . . . IBM's foundry cost is so high that it's not creating economic benefits, some companies, such as Advanced Micro Devices and National Semiconductor Corporation have given

Table 5.4 TSMC, first (and second) quarter, 2001

Item	Per cent of turnover	Per cent of orders, by factory	Per cent of orders, by area
Computer foundry applied IC	32(41)		
Foundry communication IC	31(24)		
Fabless IC design		60	
IDM		40	
USA			60
Asia			<20

Source: TSMC annual reports (www.tsmc.com/english/tsmcinfo/c0203.htm).

orders to IBM.' The fact that IBM is ahead of TSMC and UMC in process technology has made it the biggest potential competitor for domestic semiconductor factories.

TSMC has also to face the challenge from its largest competitor in HSIP, UMC (see Chapter 6). Though one senior supervisor has said that 'TSMC is just like a five-star hotel that provides superior quality of product and service and so the price will be a little bit more expensive. However, the source of customers is stable', this stability of five-star quality strategic alliances still has to rely on price tactics. In the current recession TSMC is providing its long-term customer, VIA Technologies, Inc., with a discount, yet this still does not prevent UMC from trying to make a promotion to VIA. The competition between TSMC and UMC goes back a long way. There are many differences between the two companies, such as the backgrounds of their leaders, the nature of local and foreign enterprises, and their corporate images: TSMC is a company with abundant experiences and UMC a company with lofty ambition. There is also a competitive situation between the two companies' capacities and capital. In 2000 the UMC group merged five internal companies into one company (including United Semiconductor Corp., United Integrated Circuits Corp., United Silicon Inc. and UTEK Semiconductor Corp.), with total capital of US$90 billion. Can UMC now catch up with TSMC, or even get ahead of it? Table 5.5 gives us some clues showing a summary of operating performance for the three biggest wafer foundry companies in the world, and Table 5.6 shows the competition between the two companies.

Table 5.5 Summary of operating performance (2000)

	Operating income	Gross profit	Operating profit	Profit after taxes
UMC	1050	51.0	41.5	48.3
TSMC	1662	44.5	36.4	39.2
CSM	374	33.9	13.6	21.6

Source: This study.

FROM VIRTUAL WAFER FAB TO E-FOUNDRY

Strategy Axes

TSMC has three main strategy axes – Service strategy, Technology strategy and Capacity strategy. These three axes in fact reflect the source structure of TSMC's customers. As mentioned above, the general customers of a wafer foundry can be divided into two parts. One is the so-called 'fabless' customer, the semiconductor companies of IC design without their own wafer fabs, such as VIA Technologies Inc. and Acer Laboratories Inc. (Chapters 10 and 11; see also Chapter 6). They design IC by themselves and complete the product via professional wafer foundry fabs such as TSMC and UMC. The second is the so-called 'large IDM' factory customer. They design their own IC and also produce IC products by their own fabs, such as Intel and Motorola. If they are too busy to produce their own IC products, the IDM factories will also seek the help from wafer foundry fabs. Because most of the IC designs for the orders of the large IDM factories are fixed patterns, the wafer foundry fabs can make more gross profit owing to the effects of economies of scale. The main customer source of the two biggest wafer foundry fabs, TSMC and UMC, is still concentrated in fabless customers; however, these two companies still actively fight for orders from large IDM factories.

The Service Ethic

In order to consolidate their relationships with fabless customers, TSMC puts great emphasis on 'service'. Their strategy is to give TSMC a superior impression to customers than other competitors. A wafer foundry factory with customer orientation will not only provide ordinary orders, delivery and post-sale services, but also provide a technical service to customers. Many fabless companies of semiconductor design expect that focus and resources can be centralised in IC development and design and the

Table 5.6 Summary of performance for TSMC and UMC, 2001 (NT$)

	Operating Income		Gross Profit		Proportion of orders for USA to overall turnover		Proportion of computer IC manufactured to overall turnover		Proportion of IDM order to overall turnover	
	Q1	Q2	Q1	Q2	Q1	Q2	Q1	Q2	Q1	Q2
UMC	23.6b	15.0b	40.7	15.7	46	39	25	31	28	28
TSMC	39.5b	26.3b	34.1	19.0	59	63	32	41	36	37

Source: This study.

improvement of process and cost reduction will be the responsibility of the wafer foundry factory. In order to lessen the customers' investment burden, TSMC even supplies tools and equipment for IC design to build a design-implementing platform convenient to customers. TSMC also has to make its customers aware that their products or other intellectual property (IP) will not be stolen. This is the design service that TSMC provides.

Technology Strategy

There are five main obstacles to entering the wafer foundry industry. In addition to the huge capital investment, it is also necessary to have high-quality process technology, a good customer foundation, wafer production capacity and a customer-oriented service attitude. TSMC's Technology strategy emphasises that its process technology is ahead of other competitors and, most important of all, is compatible with large international companies so that the company can establish a wide range of customers. The strategic implication of this stress on technology is that they must be equipped with high-quality production, a good rate of technology, effective cost reduction and the capacity to offer mass production.

In 2000 TSMC had the ability to provide mass production of wafer with the 0.15μm process for VIA Technologies Inc. and Nvidia. In the 0.13μm process, TSMC lags only slightly behind Intel and its plan was to get to the 0.10μm process technology by 2002. For the first quarter of 2001, the share of profit from the use of the 0.25μm process for mass production had reached more than 52 per cent. Owing to being ahead of the technology and providing sophisticated technical help to the chip-designing companies, TSMC could get a price 20 per cent higher than its competitors.

'Scale' in the wafer foundry industry means not just economic benefits but also entry barriers. In 1999 TSMC attained a market share of 37 per cent in worldwide wafer foundry markets, and UMC 33 per cent. In early 2000 market needs significantly increased, and the production capacity for wafer was well below market demand. In order to keep customers and market share, TSMC increased its market share to 47 per cent through merging TSMC-Acer and WSMC. TSMC's production capacity for 8-inch wafer has also expanded since 1999, from 1.80 to 3.46 million pieces (see Table 5.7).

Virtual Fab Services

Building on the strategic foundation of Service, Technology and Capacity, TSMC began providing virtual fab service to customers in 1997, with the intention of making customers feel as if they were producing their products in their own wafer fab, but with better technology, quality, production flexibility, response time and lower cost. To make customers think of the virtual fab as their own fab, confidentiality and information access are critical. TSMC had very early given up developing IC design and adopted a total foundry solution without trying to create its own brand and emphasising trust and loyalty to build a climate of confidentiality. By utilising information technology and electrical commerce mechanisms, TSMC set up an easy information access platform that allowed its customers to shorten the time of the design to production process. Initially the whole process took about 12–18 months, it is now down to only four months. More than 70 per cent of transactions are now dealt with by the virtual fab system (see Figure 5.2).

Table 5.7 TSMC's capacity and capacity utilisation rates 1998–2001

		Q1	Q2	Q3	Q4
2001	Capacity	1044	1084	1124	1127
	Capacity utilisation rate	70%	44%	41%	50%
2000	Capacity	680	775	930	1025
	Capacity utilisation rate	108%	101%	107%	105%
1999	Capacity	390	436	503	566
	Capacity utilisation rate	87%	102%	92%	104%
1998	Capacity	350	407	430	431
	Capacity utilisation rate	102%	75%	58%	67%

Note: 8-inch wafer in 1000 pieces.

Source: www.tsmc.com.tw.

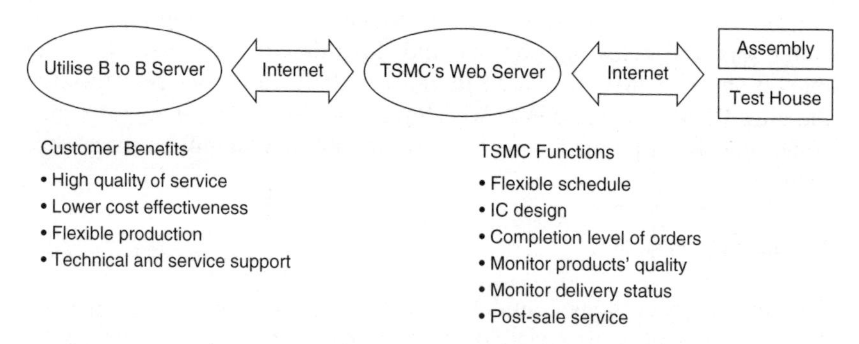

Figure 5.2 TSMC's virtual wafer service factory

The E-foundry Model

TSMC has now gone on to make its virtual fab model adaptable to growing IDM orders and achieving a faster response to customers' needs. TSMC is trying to establish an e-foundry model with powerful on-line service functions. Since customers' primary needs are to promote efficiency and increase productivity, TSMC has come up with a TSMC On-Line system satisfying customers' needs in a dynamic and e-business environment (see Figure 5.3). This system will provide customers with all the convenience and benefit of having their own wafer fabs but saves them from having to invest large amounts of capital and being involved in management issues. Customers will be able to automate the whole communication and logistic operation procedures related to technology, stock, the production process, production information, selection conditions for wafer fabs and post-sale services. Customers all over the world can enter the system through the internet at any time to query or clarify any information related to their orders.

Within the e-foundry model there is a further subsystem – TSMC-Direct (see Figure 5.4). The main purpose of this subsystem is to strengthen the relationship between TSMC and its strategic alliances. Through the integration between software and systems, mutual planning, work in progress (WIP) tracking, free-format engineering data-sharing and real-time order

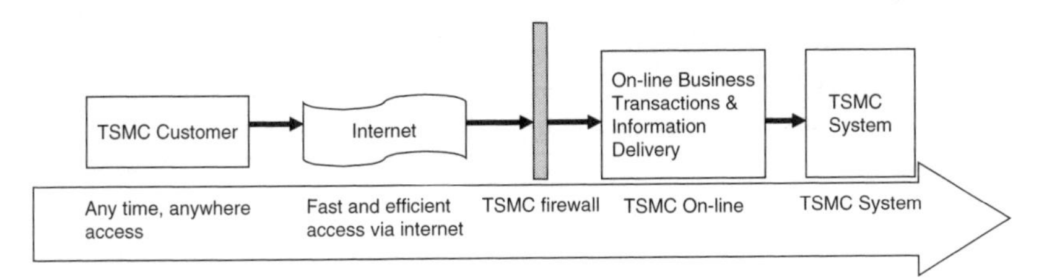

Source: www.tsmc.com.tw.

Figure 5.3 TSMC On-Line

Source: www.tsmc.com.tw.

Figure 5.4 TSMC-Direct system-to-system integration

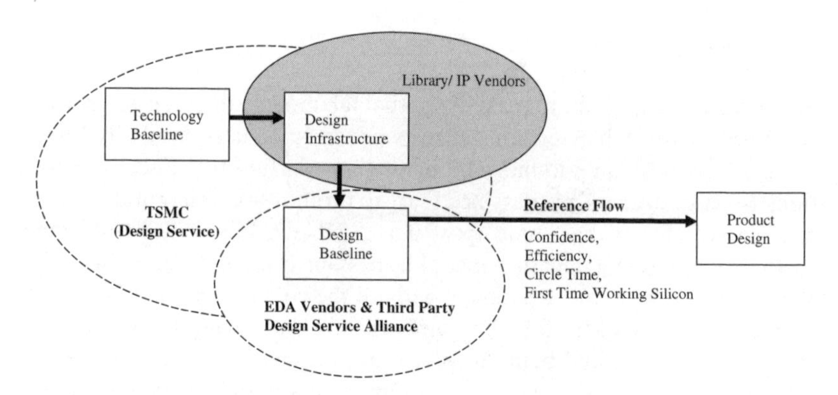

Figure 5.5 TSMC establishes the Design Service Alliance

placement can take place. The TSMC-Direct model extends the idea of virtual fab and it is expected to shorten the process cycle, cut down foundry costs and increase production quantity through such mutual participation.

IP PRODUCTS AND SERVICES

TSMC also provides a more convenient channel for verified semiconductor IP products and services. They have integrated the IC Electronic Design Automation (EDA) Vendors, Component Library, Semi-Conductor IP Vendors and product design execution service provider to establish the so-called Design Service Alliance (see Figure 5.5). TSMC's customers can give their products higher reliability and efficiency, a shorter process cycle, faster R&D and innovation through this channel. Such an integrating service will have an enormous effect on promoting customers' level of technical ability and TSMC's own added value.

TSMC'S FOUNDRY BUSINESS MODEL

TSMC's current strategy and business model is derived from their evolution of the foundry model. The semiconductor industry originated from the large IDM factories' vertically integrated model that took control of the whole process, starting from system design, wafer foundry, package/testing, to sale (see Figure 5.6). After rapid changes in the dynamic environment, the transfer and expansion of process technology and its influence on cost and scale economies, the semiconductor industry made the transition from vertical integration to vertical division of labour after 1990. In addition to

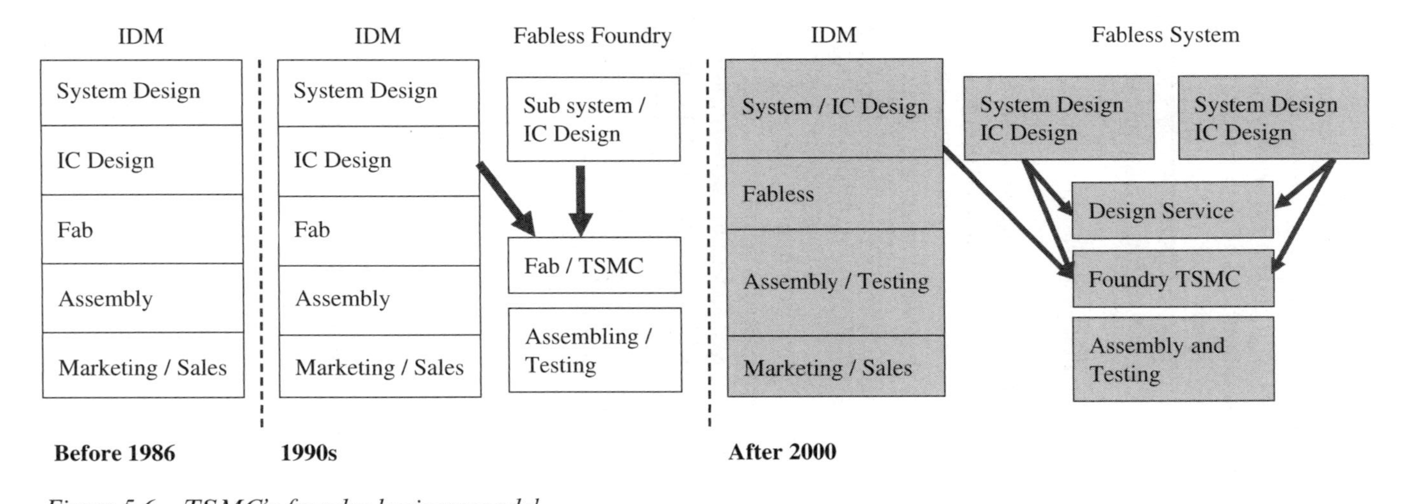

Figure 5.6 TSMC's foundry business model

large IDM factories, companies were also established for system design, IC design, wafer foundry and professional package/testing. In order to be able to continually adapt to the need to update products quickly, even the large wafer foundry companies with capital superiority had to provide design platform and tools services together with their alliance customers, and need to forge alliances with their downstream package/testing factories to ensure the performance and quality of products. The semiconductor industry in the twenty-first century has moved into a competitive era of the vertical division of labour.

Although TSMC is a professional wafer foundry fab, it still emphasises its customer-oriented service strategy. It stresses two important goals, 'leading capacity' and 'sole-source' accounts. TSMC hopes to be the wafer foundry fab with the largest capacity in the world and this 'leading capacity' includes not only the largest planning production capacity but also the lead in flexible allocation, the stable consistency of all wafer fabs' quality and the timely ability to expand. TSMC wants to be the largest wafer foundry without wafer IC design in the world and its goal is to acquire 75 per cent of all orders with wafer foundry needs ('sole-source'). For large IDM factories, TSMC expects to obtain 75 per cent of all out-sourcing orders. According to their internal assessment, they should be able to supply production capacity and service to A class customers with first priority even under conditions of limited capacity. It is above all necessary to be ahead of other competitors in process technology. TSMC's business model expects to achieve a ratio of US$1.30 of assets to US$1.00 of turnover, and to control the proportion of liabilities to assets at 30:70.

THE IMPORTANCE OF THE ENTREPRENEUR

If an enterprise is to have long-term potential and be worthy of long-term investment, the traits of the enterprise leader will be critical. Mr Morris Chang of TSMC has two traits which have helped to make TSMC a world-famous enterprise. The first is his ambition to make TSMC the number one in the world, and to realise it successfully. Second is 'leadership charm', the capacity to draw together a bunch of outstanding talents to help fulfil the enterprise's goals. Chang graduated from Harvard University, then went to MIT's graduate program, and finally attained a PhD from Stanford University. After being vice president at TI and president at GM, he accepted the invitation from Sun Yun-Suan and Li Kwoh-Ting in 1985 and came back to Taiwan to act as the board chairman of Industrial Technology Research Institute, then the board chairman of UMC. He has served as the board chairman of TSMC since it was established.

THE NEED TO CREATE A 'SERVICE CULTURE'

All business leaders, especially the founder of a company, have deep-rooted impact on the business culture. Chang told a reporter that 'a company wishing to be international should expect to be a world-famous enterprise, and the key points for success are whether the enterprise has a standardised business culture, the morality to establish an upright and honest operation, emphasises the results of long-term customer cultivation, pays attention to fostering talent, and has a strong mission sentiment from the leader downwards' (Jin, 2000). Chang thought that 'capacity', 'technology' and 'honesty' were TSMC's three most important strategic goals, and the one penetrating all of them was to provide the best service to customers. Chang above all cherishes TSMC's 'service culture' that always makes customers the first priority: TSMC will sacrifice themselves for their customers. For customers who depend 100 per cent on TSMC's wafer foundry, a failure to satisfy their needs would bring commercial ruin, making the 'service culture' the operational essence of TSMC.

An intimate interaction with customers, what Chang calls 'listening to the customer's voice', is therefore crucial. In the *Business Weekly* interview he explained that 'listening' was a key part of the communication process. Within his busy schedule more than half of his time is spent communicating with customers – in one week he had met the president of Philips, a representative of EDA Avant!, and visited an international IDM factory, National Semiconductor Corporation, and Cadence. Generally he makes contact with 30–40 CEO customers every six months. Contacting customers is strategically vital to grasp a commercial or technological opportunity or hear about company problems: 'one communication customer told me that TSMC's wafer foundry put too much emphasis on applications in the computer domain and ignored communications even when the communication market started to grow; I immediately told our vice general manager to visit the company.' The communication company became TSMC's customer and TSMC has increased its other communication customers as it successfully catches up with the new worldwide communication upsurge. 'Customers' talk makes sure I have no blind spots', Chang said. In addition to '*listening*', TSMC has set up a unique approach for interaction with customers, the Multiple Channel Contact. This is a densely covered interaction web. Chang again: 'We have about five or six channels that can be used to do cross-communication with about five or six customers, not necessarily at CEO level; sometimes it will be even between production and procurement'. TSMC's internal documents continually emphasise 'zero complaint from customers' and stress that the biggest value comes from customer satisfaction: the company regularly does customer satisfaction surveys via 300 questionnaires.

CONCLUSION

TSMC's turnover and profit have been severely affected in the late 1990s economic recession. In an environment of diminishing demand and decreasing price will it still be possible for TSMC to utilise its 'wafer foundry model' to stay ahead on technology and capacity and prepare itself for the next wave of economic growth? Since TSMC still insists on a professional wafer foundry approach with a vertical division of labour and an active 'design service', it may be able to take a relative majority of orders from fabless IC design companies, or even be a sole-source of wafer foundry services. TSMC believes that there are three key factors that will affect whether or not they can retain dominance: 'leading technologies', 'a customer-oriented service' and 'flexible manufacturing capabilities'. Is this analysis correct? Is TSMC's 'professional division of labour model' still applicable for the twenty-first century? What value will be added by the newly created 'design service'?

REFERENCE

Jin, L.P. (2000), 'The beauty and the worry of being first – how should TSMC proceed ahead? Has Morris Chang?', *Business Weekly*, **643**, www.bwnet.com.tw/.

6. Taiwan's United Microelectronics Corporation (UMC)

Soo-Hung Terence Tsai and Chang-hui Zhou

In October 1996 the United Microelectronics Corporation (UMC) held an exhibition of its IC fabrication technology in Sunnyvale, California. UMC Chairman Robert Tsao announced the establishment of the UMC Group, an 'umbrella organisation' created through the spin-off of several UMC divisions and the formation of three joint ventures (JVs) dedicated to producing ICs for companies without their own fabrication facilities. This was the latest step in Tsao's vision of making UMC the world's leading 'pure-play' semiconductor foundry.

ORIGINS OF UMC

UMC was established in 1980 as the first private-sector offspring of Taiwan's government-led strategy to foster knowledge and skills for intensive industrial development. With US$12.5 million of seed capital from the publicly funded China Development Corporation, UMC took over the government's pilot semiconductor plant and associated staff. In 1980 UMC was the first occupant of the Hsinchu Science-based Park (HSIP, see Chapter 3) which would become known as the 'Silicon Valley of the East' and Taiwan's flagship location for technology-based companies. At the time of its foundation, UMC's operations were concentrated in relatively low-tech IC testing/assembly activities for multinational customers, as well as some design work.

SEMICONDUCTORS

Integrated circuits are the backbone of all modern electronic devices. Made from polysilicon material, known by the common term 'semiconductor', ICs revolutionised many products and gave birth to entire industries. A useful way to understand the nature of an integrated circuit is to use the analogy of a road map. The basic building block of an integrated circuit is a *transistor*, analogous to a 'street'; transistors are combined to form *logic devices*, analogous to a 'neighbourhood'; logic devices are then combined to form a *chip*, analogous to a 'city'. Complex ICs contain thousands of chips and millions of transistors. Since the industry's take-off in the late 1970s, worldwide demand for semiconductors has grown rapidly, with sales increasing from US$13.1 billion in 1980 to over US$144.0 billion in 1995 (see Figure 6.1); demand is spread throughout the world – the USA accounts for about one-third of industry sales, and the rest is distributed across Japan (22 per cent), Europe (23 per cent) and the rest of the Asia-Pacific region (23 per cent) including Singapore, Korea, China, Taiwan and India. Figure 6.2 shows the industry value chain for integrated circuits.

Companies that design ICs can manufacture chips themselves, or use subcontractors. In the early 1980s the US and Japanese semiconductor firms generally manufactured their own chips, and South Korean firms (entering the industry in the mid-1980s), followed the same vertical integration strategy. After about 1988, however, the industry began to shift toward a structure where semiconductor firms began to subcontract chip production to dedicated foundry manufacturers.

FABS, WAFERS AND MASKS

Actual production of ICs occurs in an automated facility known as a 'fab', circuitry is transferred layer by layer onto silicon wafers. The total number of steps involved can reach hundreds for highly complex circuits, but the basic process involves three main operations (see Table 6.1).

IC production is one of the most complex and technically demanding manufacturing processes in the world, and the cost of building a fab has risen steadily alongside the technological complexity and scale of the production process (see Table 6.2).

The number of 'fabless' semiconductor firms grew rapidly during the 1980s. Many were small IC 'design houses' but others, such as Canada's ATI Technologies Inc., reached annual sales levels of more than US$100 million. Most of the fabless IC firms competed on the basis of high product differentiation and very rapid product cycle times. Many worked in close collaboration with their customers, often designing chips for the specific needs of a particular client application.

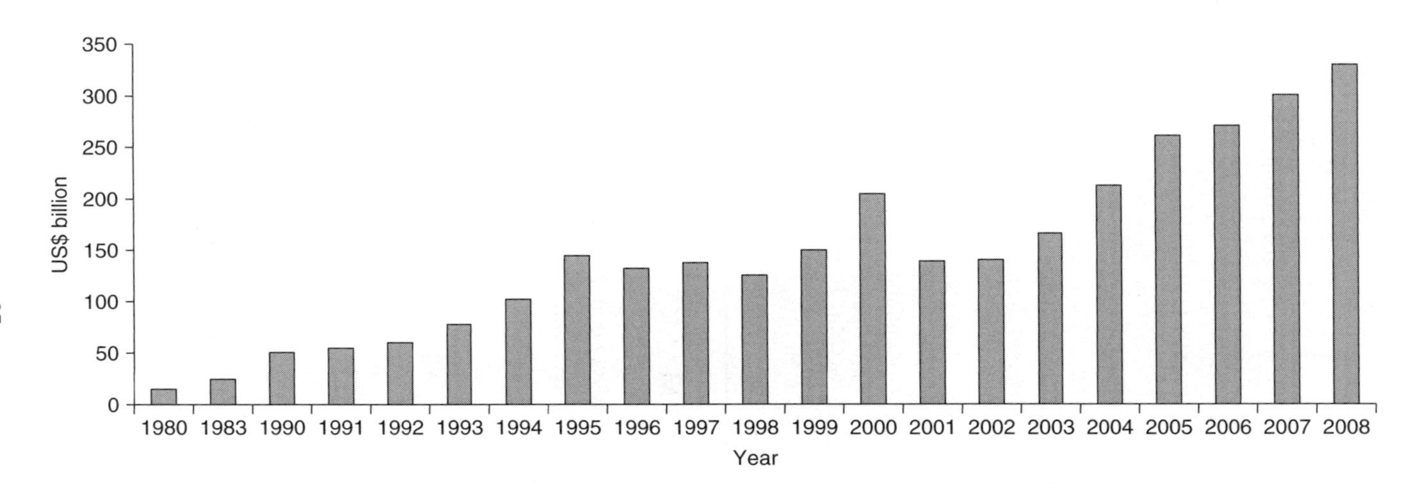

Source: Worldwide Semiconductor Trade Statistics (2005).

Figure 6.1 Worldwide semiconductor revenues, 1980–2008 (2005–08 forecast)

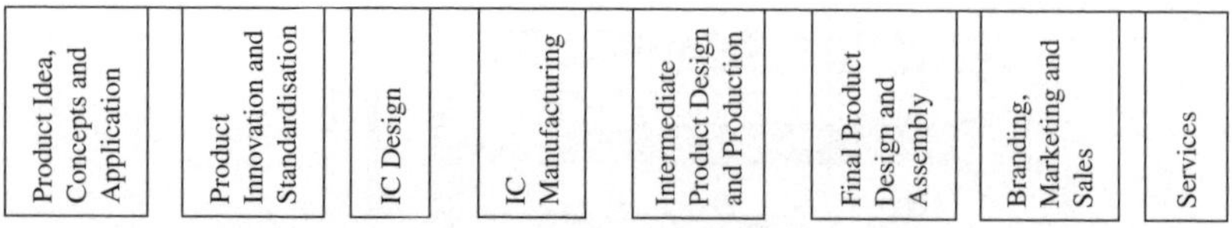

Product Idea, Concepts and Application → Product Innovation and Standardisation → IC Design → IC Manufacturing → Intermediate Product Design and Production → Final Product Design and Assembly → Branding, Marketing and Sales → Services

Figure 6.2 The industry value chain for integrated circuits

Table 6.1 The fabrication stage

- Silicon wafers are oxidised to form silicon dioxide (SiO_2), used as an *insulator* between conductors
- Selective portions of the insulator are *etched away* from the wafer using UV light, electron beams or X-rays to form patterns
- The patterns are dictated by *masks* placed over the wafer, corresponding to the design of the circuit
- This process is repeated until the entire IC is etched onto the wafer
- Each processed wafer is then doped to induce the flow of electricity
- Finished wafers are then cut into *individual unfinished ICs* before assembly and final testing

Table 6.2 The cost of building a fab

Year	Capacity	Completion cost (US$)
1983	State-of-the-art fab producing 4-inch to 5-inch wafer disks using 1.2μm line width technology[a] with a monthly production capacity of 20 000 units	250 m
1999	8-inch facility using 0.25μm technology with a monthly capacity of 30 000 units	1.5 b
2000	New fab	3 b

Note: (a) 'Line width technology' refers to the width of the transistors on the IC; the move to ever lower line widths has been driven by cost and performance factors. Lower line widths enable manufacturers to increase the number of logic devices per wafer, increase the IC operating speed and require lower voltages to operate. Industry analysts foresee a continuing trend toward larger wafers, lower line width and larger-scale fabs (see Figure 6.3).

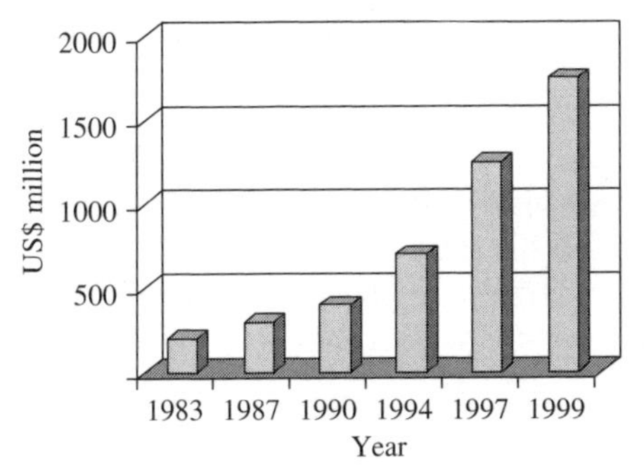

Wafer	4"/5"	5"/6"	6"	6"/8"	8"	8"
Technology	1.2μm	1.0μm	0.8μm	0.5μm	0.35μm	0.25μm
Capacity (Monthly)	20K	20K	20K	20K	25K	30K

Source: Dataquest (April 1998).

Figure 6.3 The cost of building a fab, 1983–99

THE ROAD TO A 'PURE-PLAY' FOUNDRY

Robert Tsao's Strategic Vision

In 1981 Robert Tsao was chosen from a dozen candidates to assume a vice-presidency within UMC. A year later Tsao became president, a position he held until 1995, when he was elected chairman. Under Tsao's guidance, UMC became the first Taiwan IC manufacturer to offer wafer foundry services in 1983. By the late 1980s UMC had expanded its foundry production services into the areas of memories (DRAMS) and circuits for the telecommunications industry. Tsao's vision was driven by his perspective on the technological and competitive forces shaping the semiconductor industry: it was to move UMC in the direction of a 'pure-play' foundry – a company totally dedicated to providing foundry services. In Tsao's view, mainstream semiconductor companies were increasingly focusing their strategies around systems knowledge, and IC design houses would in the future place less emphasis on pure manufacturing excellence as a basis of competitive advantage:

> IC production factors have changed a lot. Today, success in the IC industry is determined largely by how fast you keep pace and how specialised and dedicated you are at doing something, not by how many things you can do in-house.

> Technologically, the IC manufacturing process is becoming ever more complex. Manufacturing ICs is very different from designing them. I see diseconomies in doing both together. (Author's own data.)

Not everyone agreed with Tsao's vision of the forces at work in the semiconductor industry and his strategy for UMC. Company executives were concerned that the path to a pure-play foundry might be too narrow. Non-foundry operations were still making money for UMC, so why go for 'pure-play?'

DEBATES WITHIN TAIWAN

The debates within UMC also mirrored a broader debate about the future of Taiwan's semiconductor industry, especially the capacity of Taiwanese firms to move beyond 'commodity' links in high-tech value chains into more innovative business. Ferguson and Morris (1993) argued forcefully that this is the 'fatal flaw' of most Asian companies involved in high-technology sectors:

> Japanese companies, and Asian companies generally, have succeeded, by and large, by being superb commodity implementers within well-defined stable, open, non-proprietary standards . . . that are defined by regulatory agencies, other government bodies, industry standard-setting organizations, or very slow-moving incumbents . . . Non-proprietary standard products . . . memory chips, printers, VCRs, CD players, or facsimile machines, are brutally competitive businesses, with high investment requirements and razor-thin margins.

Other analysts pointed to the weaknesses of many Asian firms in developing their own brands and marketing capabilities. Few companies (Taiwan's Acer Ltd, see Chapters 10 and 11, was a notable exception) were able to break into the same 'own-brand' leagues with the major American and Japanese companies (and, to a lesser extent, the Koreans). Was contract manufacturing a viable element of the industry value chain for companies such as UMC?

UMC'S FAB CAPABILITIES

UMC steadily advanced its fabrication capabilities through a combination of licensing and internal technology development. By 1995 the company had refined its own in-house 0.5μm manufacturing process, and had followed it with 0.4μm technology. In 1996 UMC successfully launched a state-of-the-art 8-inch fab. From an original position of being 3–4 years behind the technology leaders, UMC had by 1995–96 caught up to the technological frontier. In 1985 UMC had been the world's 58th largest IC-maker; in 1996 it was 28th. In 1995 foundry revenues accounted for one-third of total revenues.

THE FOUNDRY BUSINESS IN 1996

Since its inception in the early 1980s, the foundry industry had grown rapidly. Foundries accounted for approximately 15 per cent of total IC production value worldwide, 1995 was US$135 billion. Despite a dramatic increase in foundry capacity, worldwide demand for semiconductors meant that chip demand was outrunning the capacity of foundries to supply it. The share of worldwide IC production held by foundries was predicted to increase to 25 or even 30 per cent by the late 1990s: many saw the foundry business as the next major competitive battlefield in the semiconductor industry, rather as the DRAM business had been in the 1980s.

TAIWAN FOUNDRIES

In 1995 foundries based in Taiwan produced 27 per cent of world industry output by production value. Table 6.3 shows foundry output by region, highlighting how Taiwan's share of industry output was expected to grow as existing competitors expanded capacity and new firms entered the industry: eight new companies had announced plans to build fabs in Taiwan in 1996–97 to meet the booming worldwide demand for chips. Total investment by Taiwanese IC producers was expected to reach US$16.3 billion, of which US$7.0 billion was expected to be for fab expansions or foundry startups. The balance, approximately US$9.0 billion, would go into DRAMS production. By the end of 1996 there were more than 60 IC firms in Taiwan, of which nearly 20 had fabrication capabilities.

UMC's major competitor in Taiwan was TSMC (see Chapter 5), the second private-sector spin-off (UMC being the first) from Taiwan's national initiative to create a strong and viable technology sector, especially in electronics industries (see Chapter 2). TSMC had been formed through a joint venture with the Dutch company, Philips, which transferred its advanced technology to the company. TSMC had always focused on the foundry business and by 1995 had become the world's largest 'pure-play' foundry. In 1995 revenues hit NT$31 billion; forecast sales for 1996 were in excess of NT$40 billion. Like UMC, TSMC was based in HSIP and was, in fact, the park's second-ranking occupant by sales, ahead of UMC but behind Acer. TSMC and UMC both aimed to develop the next frontier of fabrication process

Table 6.3 The world's major foundries, 1996

TSMC (Taiwan)	*Fab 1 and Fab 2:* total >100K 6-inch at full capacity *Fab 3:* running in about 20K 8-inch *Fab 4 and Fab 5:* 8-inch, starting mass production in 1997 *Water Tech:* an 8-inch US subsidiary, under construction
UMC (Taiwan)	*Fab 2:* 40K 6-inch running in full capacity *Fab 3:* 15.8K 8-inch running in full capacity Three new foundry ventures: 8-inch; USC currently running in 20K and would reach 25K by the beginning of 1997, the other two expected to start production early 1997; the full capacity for the three foundries ranged is 30–35 K
Chartered (Singapore)	*Fab 1:* running in 10K 6-inch *Fab 2:* running in 15K 8-inch
SubMicron (Thailand)	*One fab:* producing 10K 8-inch; expected 25K full capacity in late 1997, mainly to service European customers
Tower (Israel)	*One 6-inch fab:* small production
WSMC (Taiwan)	*One fab:* 8-inch, 25K in full capacity; to start groundbreaking in late 1996
IMPS (USA)	*One 5-inch fab, partially upgraded to 6-inch:* small production
Wellex (USA)	*One 5-inch fab, partially upgraded to 6-inch:* small production
GEC Plessey (UK)	*New 8-inch fab:* running in 10K

Note: Production statistics are for monthly production capacity.

Source: *Commercial Times*, Taiwan (6 October 1996).

technology (0.25μm) by 1998, and both had announced plans to move into 12-inch wafer production early in the twenty-first century.

INTERNATIONAL FOUNDRIES

The USA, Japan and Korea are the world's largest chip-making countries, but most of their major IC firms – companies such as Intel, IBM, Motorola, NEC and Samsung – utilise their own foundries.

Singapore

Competition in the foundry business otherwise came mainly from Singapore, whose Chartered Semiconductor Co. competed directly for the US customer base with UMC and TSMC. Chartered was set to launch its third fab in the first half of 1997, and had announced further plans to construct three more fabs.

Thailand and Malaysia

Two 8-inch fabs had been constructed in Thailand, one of which, SubMicron Technology, had already gone through initial production testing and was expecting to reach its full capacity of 20 000 per month by the end of 1997. Several JVs between foreign investors such as Texas Instruments and local firms in Thailand and Malaysia had also been announced.

China

Motorola Inc. had an 8-inch wafer fab in Tianjin designed to produce chips for wireless communications equipment with production scheduled for early 1997. In Shanghai, a Chinese-owned foundry operation, Advanced Semiconductor Manufacturing Co. Ltd, planned to start production at its 8-inch, 0.5μm facility at the same time.

Korea

Most mid-tier Korean firms such as LG and Hyundai Electronics Industries were still stuck in the cut-throat DRAM market and industry analysts were sceptical about their ability to enter the foundry industry on a cost-competitive basis against incumbents such as TSMC and UMC. Ramp-up times in the foundry business were long because of the complexities and scale involved in moving from fab construction to initial testing and then mass production. By the time many of the Korean firms had scaled up their 8-inch fabs, existing foundries would have already begun producing 12-inch wafers using still more advanced fabrication technologies. Many of the big Korean IC firms also carried a high debt burden, hampering the financing of new fabs.

UMC'S MOVE TO A 'PURE-PLAY' FOUNDRY

The company intended to start by spinning off various business units, including commercial ICs, memory ICs, communication ICs, multimedia

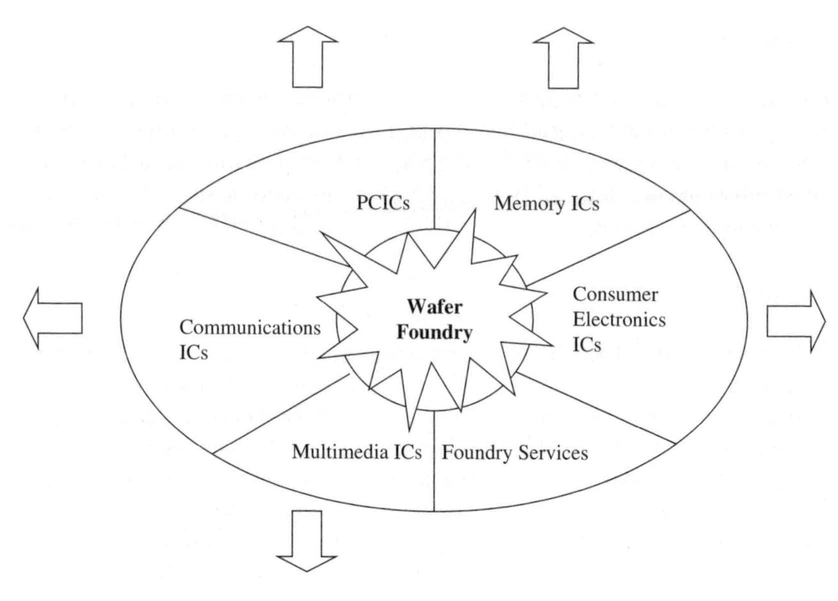

Source: UMC.

Figure 6.4 UMC's spin-off strategy

ICs, and consumer ICs (see Figure 6.4). PC ICs and communication ICs had already been spun off and established as new entities in Sunnyvale and these independent design houses were to be run by John Hsuan, the former president of UMC. The spin-off of the commercial product, memory and multimedia divisions as independent design houses based in California in 1997 would complete the restructuring programme:

> UMC will not produce its own branded IC products . . . Our fabs are just like our customers' own fabs. The difference is we do things better . . . our group structure will help ease concerns about insufficient production capacity . . . and our pure foundry concept will eliminate customer concerns about the leakage of their design ideas. (Author's own data.)

JOINT VENTURES

UMC had entered discussions with ten fabless IC design houses from the USA and Canada. In 1996 three JVs had been formed, all situated in HSIP. UMC's partnership with Xilinx, a world leader in programmable logic technology based in San Jose, illustrates the typical structure of these JVs.

In return for an investment of US$150 million, Xilinx would receive a 25 per cent stake in a new fab to be constructed by UMC. Before the joint venture fab went on-line in 1997, UMC also committed itself to meet Xilinx's current needs in wafer capacity at UMC's existing fabs. This was important to Xilinx as UMC (and other foundries) had recently had to turn down customers because of lack of foundry capacity – many fabless IC houses were scrambling to find capacity to meet the booming demand for their products.

FROM VERTICAL INTEGRATION TO DISINTEGRATION

UMC had evolved into a well-known IC player in the international semiconductor arena, successfully developing multiple IC businesses including IC product lines and foundry services. The striking issue was that, at the height of corporate development during 1995–96, UMC decided to spin-off these IC product lines and change into a pure-play foundry, giving up its own brand products and manufacturing for customers only.

For any transaction, a firm needs to commit a certain level of transaction-specific investment (site, capital, or human resources). Although the investment enhances productivity, the trade-off is that the more specialised a committed resource becomes, the lower the value in alternative uses. This is the problem of *asset specificity*. To avoid market uncertainty and organise economic activities more efficiently, firms are motivated to implement a vertical integration strategy. Harrigan (1985) has summarised the internal benefits and competitive benefits of vertical integration (see Table 6.4).

Vertical integration also raises costs, however, for several reasons (see Table 6.5; see also Balakrishnan and Wernerfelt, 1986; D'Aveni and Ravenscaft, 1994; Besanko *et al.*, 1996).

ADVANTAGES OF DISINTEGRATION

These relate primarily to economies of scale and innovation (see Table 6.6).

THE 'FABLESS MODEL'

Fabless design companies used a non-integration strategy because they were reluctant to buy specialised assets: fabs. This is especially true when the minimum efficient scale of a fab facility is large but a firm's needs are not. As suggested earlier, when adjacent stages exhibit unevenly balanced

Table 6.4 Vertical integration: benefits

Internal benefits

- *Integration economies* reduce costs by eliminating steps, reducing duplicate overhead and cutting technology-dependent costs: the economies of scope that result from technological inseparability.
- Improved *coordination of activities*, reducing inventory and other costs.
- Avoidance of *time-consuming tasks* – price shopping, communicating design details, negotiating contracts.

Competitive benefits

- Avoidance of *foreclosure* of inputs, services or markets.
- Improved *marketing or technological intelligence*. Vertical integration gives firms an improved ability to forecast cost or demand changes, thereby reducing anxieties concerning uncontrollable events. The power to guarantee supplies of raw materials or markets for products through integration strengthens firms against firms in adjacent industries. Vertical integration can provide intelligence concerning demand (especially about consumers' changing tastes), enabling firms to change product mixes quickly. Common ownership of production stages enables firms to make efficient technological adaptations at all stages so that investment expectations can be rapidly brought to fruition.
- The opportunity to create *product differentiation* (increased value-added), creating credibility for new products. Superior control of the firm's economic environment (market power). Vertical integration may also serve as an entry barrier by reducing the costs of incumbent firms and putting entrants at a disadvantage.
- *Synergies* created by skilfully coordinating vertical activities.

minimum-efficient scales, the costs of vertical integration may outweigh the benefits it yields. Obviously, IC fabrication process technology requires large volumes of throughput to be efficient. For most IC houses, internalising a fab brings management pressures (see Table 6.7).

There are therefore both pros and cons for the fabless IC design house.

UMC'S 'PURE FOUNDRY MODEL'

Why did UMC engage in vertical disintegration Keeping the IC product divisions in-house may cause three key problems. First, as foundry and IC designs rapidly progressed in their separate ways – technologies, languages, management expertise and marketing mandate – efficient coordination and control of both design divisions and foundry business will become harder and harder to achieve under the same roof. Organising expenses will

Table 6.5 Vertical integration: problems

- *Bureaucratic costs* increase with the size and the span of the firm. There is a loss of management control: time, effort, and capabilities are stretched by the need for complex administrative coordination. Managerial inefficiencies may develop because vertical integration can create complex problems of control and coordination among highly interdependent activities.
- The problem of confounding *strategic and operational issues*. Complex management tasks and operations may impede senior managers from thinking about more strategic issues.
- The *burden of excess capacity* from unevenly balanced minimum efficient-scale technology-dependent plants. Underutilised capacity may increase costs in some stages of production because throughput is unbalanced if technological factors force firms to build plants of differing scales at adjacent stages of production (think of fabless design houses and fab building). Being broadly or highly integrated may then cost more than the benefits it yields because the technologies of some plants require large volumes of throughput to be efficient.
- *Mobility (or exit) barriers* for the integrated firms. The barriers may increase strategic inflexibilities that trap firms into keeping obsolescent technologies and strategies. Firms may also be forced to forgo purchasing at low prices in the open market.

increase and the situation will become worse when foundry services become the core of UMC and its minimum-scale requirements go far beyond the needs of its own IC manufacturing.

Second, innovation problems may also emerge in UMC. IC design businesses compete on an innovation basis and this requires an *organic* structure. Foundry businesses compete primarily on high quality, low cost, fast delivery and excellent professional services, though innovation is still important. The nature of a foundry requires a more *mechanistic* structure to achieve more efficiency. The innovative quality of IC designs may be replaced more and more by an increasingly mechanistic structure. The mismatched IC product lines may also impede the growth of the foundry business because the latter need to reserve production schedules for the former at the risk of frequently interfacing with outside contracts.

Third, UMC significantly improved its IC product quality after the early 1980s, but it was still a small name, if not a 'no name' in the global arena. UMC might be able to profit from its IC products by supplying local markets, but as they become increasingly competitive it may have to step onto the global track for survival. The globalisation of the IC industry opened avenues for UMC's foundry businesses, and it found that its highly specialised foundry services could recoup high returns. Tsao had

Table 6.6 Economies of scale and innovation

- **Economies of scale.** When economies of scale are present, firms that produce more of a good or service can do so at lower average cost. Firms that use the market tend to specialise in the production of an input or output, and can consequently often achieve greater scale. A market firm can aggregate the demands of many potential buyers, whereas a vertically integrated firm will typically produce for its own use. This makes a huge strategic difference in a firm's decision about whether to target internal or external demand over the long term. Asset specificity is not necessarily bad: in the process of conducting specialised investment, firms often develop capabilities that sharpen their competitive edge. When firms or individuals make fixed investments in productive assets there is often a specialisation of production activities. It would be interesting to compare the increased benefits that the UMC as a foundry player could reap if it changed from an internal market exploiter to an external market explorer, where the external markets have been enhanced by a fabless model of atomised IC design firms.
- **Innovation**. Unlike highly integrated firms that can sometimes suffer from inefficiencies, a firm that uses markets has a stronger incentive to hold down costs and to innovate than a division performing the same activity, because it is subject to the discipline of market competition. If it fails to produce efficiently or innovate it will lose business to more efficient and innovative rivals. A division within a vertically integrated firm does not face such pressure since it usually has a captive market for its output. When common overhead or joint costs are allocated across divisions, it is often hard to measure an individual division's contribution to overall corporate profitability.

anticipated this: a foundry era was coming, which must be addressed. Keeping IC product divisions in-house would make international customers hesitate in subcontracting for the sake of private information. This, combined with the other concerns examined above, led to an inevitable decision to spin off IC product lines. Spinning off the IC lines would not only allay the fears of customers, it would also eliminate the burden of unevenly balanced scales and aggregate the demand for international IC design houses, giving room for considerable economies of scale.

THE MUTUAL 'HOLD-UP' PROBLEM

For a foundry to be successful, the fabless model is a given. By co-specialisation, each party can concentrate on sharpening its own edge and capitalise on the division of labour on a global basis. In reality things are not quite that simple. As the market takes on the role of transacting

Table 6.7 Problems of the fabless model

- Although IC design and IC manufacturing are two successive processes in the vertical chain, they are fundamentally different in their *processing nature*, requiring a different type of management attention. Most IC houses are highly innovative: fast-paced innovation is their lifeline. By internalising an IC fab, the core of innovative activities may be subject to the threat of manufacturing distraction and significant coordination costs will be incurred. When environmental conditions are turbulent – which is always true for the IC design segment – a non-integrated strategy could lower a firm's overhead and break-even point.
- As seen already, under conditions of heavy competition, integrating with fabs also increases *mobility (exit) barriers* for the IC design houses. These barriers may increase the strategic inflexibilities that can trap firms into keeping obsolescent technologies and strategies. If there are specialised foundries that are manufacturing more efficiently, keeping their own fabs may force IC houses to forgo the more attractive alternative of purchasing at low prices in the open market.
- Successful non-integration also assumes the availability of numerous *subcontractors* whose products and services equal or exceed the quality of the firm's own; if subcontractors are scarce, outsiders may gain bargaining power over the firm in a way that is disadvantageous for the firm's strategic flexibility. This is particularly true for IC design houses collectively dealing with a small number of foundries.
- Because of the relatively greater bargaining power of foundries, IC houses may have higher *transaction costs*.
- To make sure that IC products are well and smoothly fabricated, there must be a *high coordination between design and fabrication processes*. With two independent firms, especially if the foundry holds more power, it may be hard for IC design houses to achieve good coordination. This concern might be less critical because foundries always have high coordination with their customers owing to their OEM nature.
- A more critical problem is the leakage of *private information*. If the subcontractor has its own IC design divisions (as UMC did initially), the subcontractor may pirate its customers' design ideas, applications, and process technologies. Even if the subcontractor is engaged in a 'pure-play' foundry, it may transfer the information gained from one customer to another for a better deal.

mechanism, problems of *transaction cost economics* (TCE) arise. Using markets is not costless: there are the costs of gathering information on potential trading partners and drawing up agreements with them. There are then the costs of monitoring the agreement in performance and the costs of enforcement if contract obligations are not fulfilled.

For any specific contract signed with an IC house, the foundry has to adjust its fabrication process to precisely suit the design (because IC products are highly differentiated). This increases the foundry's asset specificity, which can usually be translated into sunk costs. A foundry normally requires a deposit from the IC house. This is called a 'financial hostage'. At the same time, the IC house has a responsibility to help the foundry adjust its fabrication process according to specific IC designs. The IC houses have to do so because they want to assure product quality, but during the coordination the IC houses usually use some resources (human resources, for example) to help the foundry effectively achieve the adjustment and, more critically, they have to reveal certain technological knowledge coded in the product design. In this sense, the IC firm has also committed high asset specificity for the deal. This is the situation of mutual 'hold-up': UMC and its major customers found a solution in alliances.

ALLIANCES

The formation of three JVs between UMC and its major North American customers was a move in search of a more stable equilibrium. By becoming a partner to the world's second largest foundry, an IC house like Xilinx would gain significant competitive advantage in securing supplies over its competitors that did not have such joint venturing. By bringing its customers in as equity partners, UMC would also enhance its customer base. The deal would, in effect, change both parties from short-term trading partners to long-term collaborators. Both parties would be 'going for the positive' in the sense of value creation. When trust is built along with long-term cooperation, transaction value can be achieved from:

1. Substantial knowledge exchange, including the exchange of knowledge that results in learning.
2. The combining of complementary, but scarce, resources or capabilities, which results in the joint creation of value.

Only thus can co-specialisation of both parties become a win–win game.

ENTRY BARRIERS AND SMALL-FIRM ADVANTAGES

Foundries have enjoyed 'small-number' advantages. Because foundry supply is not easily increased in the short run, foundries like UMC are able to obtain what are called 'Ricardian rents', that result from a fixed or limited supply of input factors. UMC has established a strong foundry group, enabling the company to gain such rents via alliances with its customers and also helping to enhance UMC's competitive advantages. The three JVs give UMC more production capacities and fundamentally increase its production scale. Combined with spinning off its IC divisions, UMC is able to build up its reputation as a dedicated foundry. Both scale and reputation increase entry barriers, deterring potential entrants from joining the competition.

In the IC-making industry, first-mover advantages come not only from production scale but also from the so-called 'learning curve'. Manufacturing expertise and skills achieved through learning-by-doing over many years are hard to imitate. The new fabs in China and Malaysia did not cause UMC much concern for this reason.

COMPETITIVE POSITION

TSMC was UMC's arch-rival, but the situation concerning TSMC was not simply black and white. With the existence of TSMC, Taiwan's image as America's 'manufacturing backyard' became more appealing and enhanced the overall location-specific advantages. Co-locating with TSMC could cause UMC to gain other economies of agglomeration, technological knowledge spillovers, human capital exchange and strategic monitoring/imitation. The bigger concern might be the Singapore foundries and, in particular, the Korean IC-makers that could potentially move to foundry businesses in the long run. UMC holds a better competitive position over its Singapore counterparts in production capacity and over its Korean counterparts in reputation and customer base. But these advantages may not be sustainable. To gain a sustainable competitive advantage, UMC will have to invest in firm-specific resources that are unique, less substitutable and harder to trade. Two elements are crucial for the sustainability of competitive advantages: a uniquely efficient management system and leading fabrication technologies.

CONCLUSION

By adopting a pure foundry strategy, UMC is actually specialising in activities that are to its advantage in terms of the international division of labour. What can we anticipate for UMC's future development with its 'pure-play' strategy? UMC may have underestimated the environmental constraints, particularly those related to expansion (land, human resources), industry downturn (especially downstream DRAM and computer businesses) and economic turmoil (Mathews, 1997).

NOTE

For teaching and reference purposes, a companion case 'Taiwan's United Microelectronics Corporation (UMC)' (9A98M017) can be obtained from Ivey Publishing, Richard Ivey School of Business, University of Western Ontario, Canada. The authors are particularly grateful to the Jean and Richard Ivey Fund which provided support for the research fieldwork.

REFERENCES

Balakrishnan, S. and Wernerfelt, B. (1986), 'Technical Change, Competition and Vertical Integration', *Strategic Management Journal*, **7**, 347–59.

Besanko, D., Branove, D. and Shanley, M. (1996), 'The Vertical Boundaries of the Firm', *Economics of Strategy*, New York: John Wiley.

D'Aveni, R.A. and Ravenscaft, D.J. (1994), 'Economies of Integration versus Bureaucracy Costs: Does Vertical Integration Improve Performance?', *Academy of Management Journal*, **37**, 1167–207.

Ferguson, C. and Morris, C. (1993), *Computer Wars: The Fall of IBM and the Future of Global Technology*, New York: Times Books.

Harrigan, K. (1985), *Strategic Flexibility: A Management Guide For Changing Time*, Mass.: Lexington Books.

Mathews, J.A. (1997), 'A Silicon Valley of the East: Creating Taiwan's Semiconductor Industry', *California Management Review*, **39**(4), 26–54.

7. Partner in the 'Chip Gold Rush': Applied Materials Taiwan

Tsung-yu Wu

The quality and speed of making a product depends not only on the quality of raw materials, level of production technology and human resources (HR), but also on good production equipment and tools. The adage 'You cannot make bricks without straw' holds particularly true in the semiconductor industry. During the semiconductor production processes – from disposition, photolithography, etching, ion implantation, to polishing and cleaning – the smooth completion of each step requires well-developed equipment and services.

THE RELATIONSHIP BETWEEN SEMICONDUCTOR MANUFACTURERS AND EQUIPMENT PROVIDERS

There exists a mutual relationship between semiconductor manufacturers and equipment providers. Equipment providers are important backstage participants in semiconductor manufacturing. In accordance with Moore's Law, the number of transistors that the makers of semiconductors are able to put on a chip doubles every two years. Since it was proposed in 1965, this law has so far held true. Since the efficiency of semiconductors is doubling every two years, the implication is that chip-makers will be knocked out of the race if they do not keep up with the most advanced technology. The ability of chip-makers to produce smaller, faster, and more effective chips is being determined by the sophistication of their production facilities. If production facilities fail to keep up with the rapid advances in technology, even chip-makers with the strongest financial foundation, the best production technology and the most talented engineers will still be crushed by the competition. Chip-makers are also the most important customers of facility providers. Take investment in 8-inch wafer plants as an example: the total amount of investment in Taiwan's semiconductor industry is NT$25 billion, 70 per cent of it earmarked for the purchase of facilities. Taiwan's semiconductor industry spent US$9 billion for manufacturing equipment in 2000, accounting for 20 per cent of global sales and reflecting a 106 per cent growth from 1999. Because of the close relationship between chip-makers and facility providers, the role of all partners in the 'chip gold rush' must be taken into consideration to get the whole picture of the development of both Taiwan's and the global semiconductor industry.

APPLIED MATERIALS INC.

Applied Materials Inc. is a multinational company providing chip-makers with production facilities, technology and services. Founded in 1967, Applied Materials has become the world's largest equipment provider for wafer production and the most important partner for chip-makers. Its products include many kinds of semiconductor production equipment, such as chemical vapour deposition (CVD), physical vapour deposition (PVD), epitaxy and polysilicon vapour deposition, rapid thermal processing (RTP), ion plantation, etching, process diagnostics and control (PDC), chemical machine polishing (CMP), mask production, flat panel display (FPD), and manufacturing execution system (MES) software. Applied Materials is headquartered in Santa Clara in the United States, operates several technology development centres in America, Japan, Europe, Israel, Korea and Taiwan, and has set up more than 90 sales and service stations in 13 countries in America, Europe, Israel, Japan, Korea, Taiwan, Singapore, Malaysia and China. All branches are located near main clients or in the semiconductor manufacturing centres of each country, with over 17 000 employees around the world. Its annual revenues were over US$1 billion in 1993 and over US$5 billion in 1999. It has been the established leader in products such as CVD, PVD, RTP, etcher and CMP for many years (Applied Materials Inc., 2000).

AMT

Applied Materials Taiwan (AMT), headquartered in the Hsinchu Science-based Industry Park (HSIP, see Chapter 3), is one of the subsidiaries of Applied Materials Inc. Established in 1989, AMT took only ten years to become the largest semiconductor manufacturing equipment provider in Taiwan. Its annual revenue in 1999 was 21 per cent of the global sales revenue of Applied Materials, behind only the North America operation (34 per cent). AMT has over 30 customers, including TSMC, UMC, Winbond, Mosel, Macronix, NTC (Nanya Technology Corporation), Powerchip (Powerchip Semiconductor Corporation), and Holtek (Holtek Semiconductor Inc.). AMT also has sales and service centres in Linko and

Table 7.1 Factors in AMT's success

- Support of its parent company
- Upsurge of Taiwan's semiconductor industry
- Background and leadership style of AMT's founder
- Customer-oriented management philosophy of the firm
- Spirit of continuous R&D
- Achievement of first-mover advantage
- Comprehensive management system
- Implementation of localisation

Tainan, a technology centre established in 1996 and an Asian Continental Distribution Centre in Taoyuan, set up at the end of 2001 to expand its service to every corner of Asia (AMT, 1998; AMT, 2001).

How could AMT become such an important player in the development of Taiwan's semiconductor industry so rapidly? Was this the result of its parent company's success, or of AMT's own efforts? Was AMT's success the result of Taiwan's success, or was Taiwan's success in developing a semiconductor industry the result of AMT's success? This chapter attempts to analyse eight major factors contributing to AMT's success (see Table 7.1).

APPLIED MATERIALS' DEVELOPMENT

AMT's rapid expansion can be partly attributed to Applied Materials' success. With the advantages of plentiful capital, complete product lines, continuously advancing technologies, macro strategies of globalisation, a good social image and mature strategies for market entry, Applied Materials was able rapidly to obtain a competitive advantage in Taiwan and around the world. There were several turning points in the process of becoming the leader of the industry. On its foundation in 1967, just when the semiconductor industry was beginning to develop, the company had five employees and funding of about US$100 000. At that time, chip-makers tended to research and develop their own manufacturing equipment, so Applied Materials' main business was to provide them with the materials and components necessary for equipment production. Equipment made by chip-makers often lacked consistency and reliability, however, so Applied Materials started to provide 'turn-key' chip-making equipment, assuring its customers that as a result of rigid testing in the laboratory it would meet their requirements. Applied Materials entered the field of semiconductor manufacturing equipment.

Applied Materials' success thus came about through the intersection of a growing chip industry, its specialisation in value-added equipment production and a customer-oriented spirit. By 1972 Applied Materials had sold equipment to a wide range of semiconductor manufacturers around the world; with sales revenues of US$6.3 billion and 155 employees, it went public on the stock exchange. The rapid growth of the company and continuous expansion of product lines, however, led to overextension of resources and financial crisis. James C. Morgan, who was named president and CEO in 1977, looked at the situation and defined semiconductor manufacturing equipment as Applied Materials' core business. This decision changed the company's product strategy from breadth to depth, and its corporate strategy from diversification to clear focus. Providing semiconductor manufacturing equipment and services became Applied Materials' core goal.

Since 1980, Dr Dan Maydan from the world-famous AT&T Bell laboratories, Dr Sass Somekh and Dr David K.N. Wang all joined Applied Materials, helping the company to research, develop and commercialise many innovative technologies. These activities solidified Applied Materials' critical strategy of 'continuing research and development of products', which became an essential part of the company's organisational culture. In 1987, Applied Materials introduced to the market the first product in the Precision 5000 series featuring a single-wafer and multi-chamber platform in which four key processes are integrated to increase the efficiency of chip production. Because of the throughput improvement this technological innovation brought to semiconductor manufacturing equipment, Precision 5000 was hailed as 'the most successful product introduction in the history of the semiconductor equipment industry' (author's own data): in 1993, the Smithsonian Institution placed the first Precision 5000 system in the permanent collection of its 'Information Age' exhibits in recognition of its importance in the history of the semiconductor industry.

Applied Materials' annual revenue was over US$500 million in 1989, and it broke into the ranks of the Fortune 500 in 1990. With the establishment of the Austin Volume Manufacturing Centre, the company became the world's largest semiconductor equipment manufacturer. *Fortune* magazine recognised Applied Materials as one of America's most admired companies as well as one of the 50 best places to work. *Industry Week* rated Applied Materials as one of the world's 100 best-managed companies. Applied Materials is now the top company in the semiconductor equipment industry, leading other companies in areas such as sales, services, production and technological development, and steadily expanding its overseas markets without encountering many obstacles (Applied Materials Inc., 2000; AMT, 2001).

Applied Materials has historically been able to foresee opportunities and make prompt decisions when facing crises, environmental changes and threats. In addition, Applied Materials gained competitive advantage by making efforts to satisfy customers' needs and continuously innovating and developing new technologies. The company philosophy and spirit not only led to Applied Materials' success in the United States, but also helped its branches and subsidiaries around the world to fuel their success. Applied Materials was able to successfully lead its branches and subsidiaries not only through concrete support (such as finance, products and technology), but also the transference and inheritance of its corporate spirit.

JUMPING ON TAIWAN'S SEMICONDUCTOR TRAIN

As part of its globalisation strategy, Applied Materials actively enters overseas markets whenever it discovers a niche. In 1977 Applied Materials foresaw a potential market in Europe, and set up Applied Materials, Europe (AME) to handle its business there. In 1979 and 1984, Applied Materials similarly expanded into Japan and China and built Applied Materials, Japan (AMJ) and a service site in Beijing in order to capitalise on a rapidly expanding market. Applied Materials later followed the same path in Korea, Taiwan and Singapore by establishing Applied Materials, Korea (AMK), Applied Materials, Taiwan (AMT) and Applied Materials, South East Asia (AMSEA) as regional subsidiaries. All these regional units were established and developed concurrently with the local semiconductor industry.

AMT IN TAIWAN

Applied Materials' major business in Taiwan was originally to import equipment and resell it locally. Yet, as clients continued to have technical problems and available services and support were insufficient, more and more voices suggested that the company should set up an official site in Taiwan for professional supervision and assistance. With the Taiwan government's support, more and more privately owned chip-makers were established in the mid-1980s, and the establishment of foundry plants such as TSMC and UMC (see Chapters 5 and 6) also proved Taiwan's abilities in chip production. A niche market clearly existed for semiconductor manufacturing equipment in Taiwan. Chiam Wu, an etching manager in Applied Materials Inc., therefore made the important decision to expand to Taiwan. Wu initially came back to Taiwan in 1989, to handle TSMC's technical etching problems, but then asked for a transfer to Taiwan to set up a new site. AMT was established at the end of 1989, in charge of sales and services in the Taiwan area. This might have seemed a risky decision, since there were only six wafer manufacturers in Taiwan by 1989, and revenue totalled only US$224 million. Yet some large chip companies such as TSMC and UMC were just beginning to turn a profit and were braced for rapid growth. Everyone had confidence in the nascent industry, but still felt uncertain about its prospects.

Wu's decision was proved correct. The government began to advance the 'submicron project' in 1990 and succeeded ahead of schedule in 1994. Taiwan chip-makers were then able to use 8-inch wafer manufacturing and DRAM and SRAM development technology, putting the Taiwan semiconductor industry on the world's stage as a hi-tech manufacturing centre to be respected and feared by developed countries. AMT jumped on the express train of Taiwan's semiconductor industry and rapidly grew just as the industry gathered speed. In 1991 AMT's sales and orders amounted to only US$24 million and US$26 million, respectively. In 1993 new orders broke US$40 million and total installations exceeded 100 systems. AMT changed its status from a branch to a subsidiary in 1993, and Wu was promoted from sales and marketing manager for the Taiwan area to the general manager of AMT. By 1994, total annual revenues for Taiwan's semiconductor industry had exceeded US$2 billion, almost ten times that of 1989, and AMT received orders worth more than US$100 million.

AMT's sales growth was in line with Wu's expectations, but its speed was far beyond what she had expected. AMT was not able to deal with customers' technical requirements using its management strategy of machine sales, so in 1994 Wu made the second important decision to found a 'Technology Centre' in Taiwan. This was another dangerous decision given that AMT's annual revenue was only about NT$800 million, while the investment in the centre would be over NT$1 billion (about US$30 million). The centre, located in HSIP, began construction in 1994, and was open for business in 1996. With an application laboratory, manufacturing centre and training centre, AMT had stronger abilities in researching, developing and improving machine functionality. AMT also started cooperative projects with customers and developed products such as the Wafer Lift in 1997, quartz parts in 1998 and PVD Degas in 1999. These achievements again proved AMT's R&D ability to the parent company and the Taiwan semiconductor industry as a whole.

AMT AS A DRIVING FORCE IN THE SEMICONDUCTOR INDUSTRY

Wu's vision of the future became reality. In 1995 the Taiwan semiconductor industry entered a new era. The volume production of 8-inch wafers

Table 7.2 AMT: installed bases and orders, 1996–2000

Year	Installed bases	Orders (US$)
1996	500+	400 million
1997	800+	1 billion
2000	2000+	2 billion

Source: Applied Materials Taiwan (2001a).

and potential for 12-inch wafer production showed that Taiwan's semiconductor industry had matured. More and more companies began investing in wafer foundry plants. Since each chip-maker possessed sufficient R&D ability, the capability to provide manufacturing equipment that kept up with chip-makers' R&D progress became one of the most critical factors for success. The establishment of AMT's Technology Centre met customers' urgent needs (see Table 7.2).

In 2000 Taiwan's semiconductor industry revenue totalled US$15 billion, over 60 times revenue in 1989, and seemed to have limitless room for growth. Because of its precise understanding of the development potential of the Taiwan market and the strong connection between AMT's operational strategies and the development of Taiwan's semiconductor industry, AMT had obtained a leading edge. AMT witnessed the development of, and became one of the driving forces of, Taiwan's semiconductor industry.

CHIAM WU: THE SOUL OF AMT

Educational Background and Career

Chiam Wu played a critical role in AMT's transition from a pure sales site to the company of today. There are many female general managers working for foreign companies in Taiwan. However Wu is the only one to hold the position in such a large high-tech foreign company for such a long time, in a field in which males are in the majority. What are the secrets of Wu's success?

Wu grew up in the countryside in Taiwan and received her secondary education in the Wesley Girls High School. After graduating from the Department of Materials Science of National Tsing Hua University in 1978, she went to Oregon University to study towards her Master's degree in Materials Science, which she achieved in 1980. She first took a position at Signetics as a production process engineer, and in 1982 transferred to Advanced Micro Devices as a senior engineer. In 1987 she became manager of the etching application laboratory at Applied Materials Inc. In 1989 she went back to her home town in Taiwan and started a new career (Chuang, 2001).

In AMT and other hi-tech companies in HSIP, a number of high-level managers had a career path similar to Wu's. They grew up and finished their fundamental education in Chinese society, went to the United States to study towards advanced degrees and work, and later returned to Taiwan to embark on the second leg of their career (Addison, 2001). Take the scientist Irving Ho as an example; he was part of the first graduating class from the Department of Electronic Engineering in Xiamen University, and later moved to Taiwan with the government in 1949. In 1956 Ho went to the United States and received his Master's and doctoral degrees from Stanford University. He worked in an IBM laboratory for 16 years and obtained 34 American patents. In 1979 he went back to Taiwan and became the first manager of HSIP. As another example, Morris Chang was born in Ningbo, Zhejiang, and spent most of his youth in Mainland China. He went to the United States and received his Bachelor's and Master's degrees in the department of mechanical engineering at MIT, as well as his doctoral degree from the department of electronic engineering at Stanford University. After being a vice-president at Texas Instruments and General Instruments, Chang went back to Taiwan and founded the Taiwan Semiconductor Manufacturing Company (TSMC) in 1986 (see Chapter 5), becoming its president. Minqui Wu, the general manager of Macronix International Company (MXIC) (see Chapter 4), graduated from National Cheng-Kung University and received his Master's degree from Stanford. He was an engineer at Intel and a manager at VLSI Technology, and went back to Taiwan in the early 1980s to establish MXIC. Within AMT, high-level managers such as President David Wang, Vice-CEO Bonin Chang and Vice-General Manager of the Product Department Chinjung Wang all have similar backgrounds. Entrepreneurs with similar career paths establishing and developing hi-tech companies usually have five key characteristics in common (see Table 7.3).

Interpersonal Relationships

Yet a good personal career background is no guarantee of company success. Wu's special role in the Applied Materials' system was also an important factor in AMT's achievement. AMT's outstanding performance gained much appreciation from the parent company and this gave Wu a firm position from which to fight for more resources and autonomy: AMT's transition from a branch to a subsidiary in 1994 and the establishment of the Technology Centre in 1996 are two key examples. Wu was promoted to

Table 7.3 Key characteristics of high-tech entrepreneurs

- Have **international perspective**, understand and identify with the local industrial environment
- Have had **experience in American hi-tech industries** and have good professional and technical knowledge; they can therefore easily apply their personal specialties learned in the United States to competitive advantage in establishing companies in Taiwan
- Have **strong social networks** in Japan and America, which help them to expand their business
- Are **action-oriented** and ambitious
- Have taken **similar paths** in career development, education, entrepreneurship experience, common language and lifestyle; they become close partners in business and provide each other with encouragement in expanding their own businesses

Applied Materials' vice-president in 1996, global vice-president in 1998, and global Group vice-president in 1999, pushing her to the core of Applied Materials' global management and making her more influential at the highest level of decision-making in the company. Wu was also good at establishing interpersonal networks with high-level leaders of other companies and with customers, both business and personal: many companies in HSIP now are AMT's loyal customers.

Strategic Planning and Management Philosophy

The *Economics Daily* described Wu's management philosophy as 'unique style, delicate manners'. Wu accurately predicted the development trends of the semiconductor industry and proposed appropriate strategies to respond. AMT's manufacturing centre, for example, became the pioneer of Applied Materials' overseas manufacturers after 1994. In 1996 AMT Technology Centre was established to strengthen cooperation with customers in R&D. In 2000 AMT actively collaborated with the government to advance the 'flowing water' project (see below) and established a complete supply chain for the local semiconductor industry, a project that received accolades from industrial communities. Wu brought the parent company's strategies of customer orientation and continuing R&D to Taiwan and made efforts to improve production quality (see Table 7.4). AMT has achieved better performance on many dimensions and received more quality certificates than its parent company.

Wu has also proposed a number of employee welfare measures and development strategies, such as employee stock options, three-year

Table 7.4 AMT's quality awards 1996–2000

- ISO9001, ISO14001, and OHSAS18001 certificates
- Industrial Excellence Award (MOEA)
- Industrial Technology Advancement Award (MOEA)
- Excellence Quality Management Award (MOEA)
- R&D Engagement Award (HSIP)

Source: Applied Materials Taiwan (2001a).

interest-free auto loans, free breakfast, holiday and vacation policies exceeding labour law requirements, a luxurious fitness centre, 'green hill' plans (see below) to improve employees' mental and physical health and encourage their lifetime learning, and a complete HR development system, making AMT one of the best companies in HSIP and helping it to attract talent.

On 30 April 2000 Wu received the First Outstanding Alumni Award from Dr Liu, the president of National Tsing Hua University. Her success is typical of Taiwan hi-tech semiconductor founders.

AMT'S CUSTOMER-ORIENTED STRATEGY

Customer business results have always been AMT's highest priority. AMT's mission is 'through systems, manufacturing process models, and services, to continue innovating products and improving customers' manufacturing ability in the hope of becoming the leader of the world's semiconductor manufacturing equipment and service providers' (company's internal data; author's own data). AMT considers customers as its first priority. Its core values include: providing products and services needed to its customers, listening to customers' opinions systematically and responding in a timely fashion to customers' requests. AMT claims that '82 per cent of their human resources engage in customer-related services' (company's internal data; author's own data).

Customer Service Centre

In order to facilitate more direct customer contact, AMT set up a 'customer service centre', consisting of several account management teams (see below). Led by a manager, each team contains a group of sales representatives and technical service people, and is responsible for one or more clients with their existence and success tied directly to the accounts. Two separate teams were in charge of TSMC and Texas Instruments–Acer Inc., but when Texas Instruments–Acer merged with TSMC, the team dedicated to it was also merged with the TSMC team.

Account Management Teams

Among the global Applied Materials' entities, AMT was the first subsidiary company to use 'account management teams' to run their business (Yen and Liu, 1996). This kind of organisational design has five key advantages (see Table 7.5).

Product Technology Administration

AMT has a product technology administration to provide support for customers' manufacturing processes and software technology. The difference between a product technology administration and a customer service centre is that the former is based on products and the latter is based on customers. More specifically, the product technology administration is responsible for the integration of manufacturing and R&D in technology, while customer service centres are in charge of easy machine maintenance and repair. In the product technology administration, each kind of product has a group of experienced engineers and researchers who work together as a team to coordinate and provide customers with the newest product technology. The major goal is to resolve problems in customers' manufacturing processes in order for them to achieve the best production volume possible.

Customer Productivity Support System

To increase the added value of products and services, AMT set up a 'customer productivity support' system in which three key services were provided (see Table 7.6).

Customer Satisfaction Department

In order to improve customer satisfaction with Applied Materials' products and services and to build a basis for improvement, AMT set up a customer satisfaction department in charge of regular customer satisfaction investigations, focusing on their needs, expectations and satisfaction with AMT's services. Four kinds of surveys are included in AMT's customer satisfaction investigations (see Table 7.7).

Table 7.5 'Account management teams' at AMT

- Each team has both **sales** and **service people** so that both sides can support each other and avoid problems arising from a lack of cooperation between them.
- In hi-tech companies, sales and technical service personnel are usually in different departments. When the two departments grow independently, it is common to see a lack of cooperation grow between them. This conflict is mostly due to differing performance measures among the two groups. Sales people are judged by the number of machines they sell, and gain bonuses dependent on their sales performance: thus, from the salesperson's point of view, the more orders they obtain, the better. On the other hand, workers in technical support usually receive fixed salaries, and thus their incomes do not vary according to the services they provide: from the point of view of technical service people, the less support they are required to provide the better. When salespeople drive too many orders, conflict can arise with technical people who refuse to handle the new cases.
- This can seriously damage the quality of customer service provided. AMT adopted an organisational design that merges sales and service people into one unit so that both can face the client together and foster mutual responsibility. This design improves cooperation between sales- and service people; customers also obtain solutions and feedback rapidly regarding their unique problems.
- If sales and technical people are responsible for all customers, customers may worry about confidentiality issues. When a technical person goes to UMC for machine maintenance right after finishing his service at TSMC, what guarantee is there that he/she will not reveal confidential information about TSMC to UMC? Likewise, would a salesperson divulge information about a client in order to secure orders from a competitor?
- Even though these questions are hypothetical, they still affect customer trust in AMT. To shore up customer confidence, AMT thus adopted the strategy of 'account management teams', assigning only one or two customers to a particular team. For some critical clients such as TSMC and UMC, AMT splits up team responsibility so that different teams are in charge of different accounts. In so doing, AMT can reassure their customers and improve satisfaction. Customers can experience one-to-one attention, which not only raises the effectiveness and speed of service, but also **makes customers feel valued**. This fits in very well with Applied Materials' spirit of customer orientation.

Table 7.6 AMT's customer productivity support system

- **Spare parts service** In semiconductor production, all activities stop when there is damage or a shortage in equipment components. Receiving components of good quality in time thus becomes one of the most important issues. Yet this brings equipment manufacturers difficulties, given that the components needed are not always easily delivered to clients, and component storage and management are resource-consuming. AMT proposed spare parts service packages: customers place orders based on their own estimated need for parts; based on customer estimates for the coming 12 months, AMT decides on annual purchases and replenishes stocks so as to minimise storage costs. AMT also holds meetings with each customer on a monthly base to examine the trend of spare part use in order to improve stock management and purchasing accuracy. AMT is also fully responsible for parts planning related to the maintenance of customer manufacturing equipment. This integrated support package (ISP) includes AMT's commitment to provide and prepare parts in time, assist customers and other providers in the management of spare parts, assign engineers to be on standby and in charge of manufacturing equipment repair and regular maintenance around the clock. This package significantly reduces customers' costs in equipment maintenance and spare parts management.
- **Product refurbishment and enhancement (retrofitting)** Product refurbishment can keep equipment in good condition, while product enhancement improves machine functioning, extends machine life, and increases machine throughput. Product refurbishment services include whether the customer chooses to utilise such services at their own plant or outside, a 90-day quality warranty and assurances that the results meet system and safety requirements, and is indistinguishable from new machines. As part of its product enhancement services, AMT provides *product improvement kits* to upgrade parts and components of customers' manufacturing equipment (such as upgrading HP machine arms to VHP machine arms), *configuration upgrades* to upgrade current manufacturing equipment standards (such as upgrading reaction chambers from Mxp to Super E), and *tool re-configuration* between different production processes. AMT even assists customers in moving manufacturing equipment and providing all kinds of services.
- **Technical training** After the inauguration of the technological R&D centre, AMT's local technical training centre also came into use. Customers were able to receive training courses in HSIP, without having to go abroad, avoiding problems of technical gaps, language barriers and cultural misunderstandings. This training centre also gave unexpected benefits, such as cutting customer training costs, improving training effects, fostering specific technologies and improving customer satisfaction. The training centre had by 2000 delivered training courses to over 10 000 people in the industry.

Source: Applied Materials Taiwan (2001b).

Table 7.7 AMT's customer satisfaction surveys

- **Annual Customer Relation Survey (ACRS)** ACRS is conducted annually by a consulting firm. Its purpose is to monitor worldwide customer satisfaction with and loyalty to Applied Materials' and competitor products through phone interviews.
- **Installed Base Support Survey (IBSS)** IBSS is also conducted annually by a consulting firm. Its purpose is to monitor worldwide customer satisfaction with and loyalty to Applied Materials' and competitor support/services, technical training and parts provision through phone interviews.
- **Executive Listening Session (ELS)** This is conducted twice annually. Local account general managers hold face-to-face interviews with high-level managers of client companies.
- **Individual Product Satisfaction Survey** This is conducted quarterly by a local department on customer satisfaction and focuses on different products each time. All parts of the survey data are recorded in AMT's database of customer satisfaction so that related departments can set up improvement plans and periodically track the improvement results.

Source: Internal documents about Customer Satisfaction Survey (1999). Applied Materials Taiwan (2001b).

AMT's 'Total Solution'

'Total solution' is a logo that can be seen everywhere in AMT. It represents AMT's determination to 'respond to customers' needs and execute strategies that satisfy customers' (author's own data). AMT's total solution in practice contains three parts:

- Being close to customers
- Researching and developing advanced semiconductor manufacturing processes
- Achieving world-class team performance

Being close to customers is ranked first. Through its design of account management teams, planning of product technology administration, abundant customer productivity support and customer satisfaction surveys, AMT embodies the parent company's spirit of 'customer orientation'. AMT has also implemented some unique measures such as the 'account management teams' and a local technological training centre to highlight the importance of localisation. These have been the critical factors in AMT's success in Taiwan (AMT, 2001b).

THE SPIRIT OF CONTINUING R&D

Commitment to R&D

Throughout AMT's history, important turning points always involved issues of R&D. Applied Materials invited famous scientists from AT&T Bell laboratory to join its team in the 1980s, and introduced some important products such as the Precision 5000 series, Endura PVD and RTP Centura to the market in 1987, 1990 and 1995, respectively. It also established the Equipment Process and Integration Centre in 1998, and started to develop technology enabling a new copper production process. These events helped Applied Materials to achieve some critical early successes. For a decade, Applied Materials on average contributed 13.8 per cent of its revenue to R&D, and endeavoured to develop more effective and functional machines, software, and production processes in order to improve production efficiency and increase production of semiconductors. Continuing R&D helped Applied Materials expand its product lines to cover all kinds of equipment needed in the semiconductor manufacturing process. High-level technological support and strong process integration gave it an even higher level of competitive advantage compared with competitors with fewer product lines.

Priority of Research

AMT made research a priority. Yet compared with their parent company, most subsidiary companies of Applied Materials have less technical ability in R&D and advanced product production and can engage only in selling machines and providing technical support. Thus when clients request higher-level technical assistance, subsidiary companies are usually unable to respond on their own and must wait for support from Applied Materials Inc. The long distance between local clients and US offices often slows delivery of support and services. Cultural and technical barriers also degrade the quality of service and competitiveness. Subsidiary companies' growth was also restricted because the America-based parent company continuously restrained their technological development. The most important strategy in breaking this bottleneck was to improve subsidiary companies' R&D ability.

AMT clearly saw that it was not enough to satisfy local needs solely through technical support and R&D in America. The only way to consolidate its position in Taiwan, win the respect of the parent company and in turn strive for more autonomy was to develop independently its own R&D abilities, obtain customer recognition and attract more orders. As Chiam Wu said:

> At the time of AMT's establishment, customers requested only hardware support, and we provided simply sales and services. Later, customers expressed the view that it was inefficient and time-consuming to ship components from foreign countries, and thus hoped AMT could provide components from local suppliers. Customers also wanted AMT to work with them on a more intimate basis to provide support tailored to their individual needs, as well as to research and develop suitable equipment. Thus, as customers' production processes evolved, we also needed to adjust ourselves to meet their new support needs. (Yen and Liu, 1996, pp. 37–8)

In line with increased customer requirements, AMT started to develop its own technological support system and research and development ability to keep those customers.

Technological R&D Centre

To satisfy customer requirements in technological support and R&D, AMT established a 'technological R&D centre' in HSIP. The centre helped to facilitate just-in-time (JIT) technological support for AMT's local clients, and helped decrease customer dissatisfaction with the parent company's tardy support. AMT also proposed a 'continuous improvement project' (CIP): when clients have problems with equipment throughput or part quality, local offices are empowered to solve the problem, rather than having to rely on parent company assistance. To enhance cooperation with clients to develop more sophisticated equipment and promote production process integration, AMT then proposed the 'Joint Development Project' (JDP), through which AMT actively develops partnerships with customers and makes efforts to decrease costs, increase output and improve semiconductor manufacturing technology.

AMT as an 'Active Provider'

AMT attempted to develop its own production ability in order to switch its role from a 'passive demander' to an 'active provider'. In 1994 AMT set up its own manufacturing centre and became a manufacturing pioneer among Applied Materials' branches and subsidiaries (see Table 7.8).

'Developing advanced semiconductor production ability' was another way for AMT to carry out its philosophy of providing a 'total solution'. AMT's efforts in solving customer problems, cooperating with customers in R&D and actively manufacturing its own products helped it to meet its goal of continuous R&D, and became a critical factor in its success.

Table 7.8 AMT as manufacturing pioneer

Year	Development
1995	Researching, developing and repairing RF-Match in the local manufacturing centre to meet global customer and local client needs by cutting the time of product shipping.
1997	Mass-producing Wafer Lift in Taiwan and providing it worldwide.
1998	Developing, mass-producing and providing to chip-makers globally critical parts of semiconductor manufacturing equipment, such as DPS RF-Match, PVD, HP Wafer Lift and PVD Degas. Both P5000 Wafer Lift and RF-Match received ISO9000 certification.
1999	Succeeded in developing the Enhanced PVD Degas.
2000	Started local development of 300 mm and copper production process equipment and became the first overseas branch area authorised to execute this production project outside the United States.
2001	Started to produce Orientor Chambers and Cooldown Chambers in Clean Rooms to provide to chip-makers worldwide.

Source: Applied Materials Taiwan (2001d).

AMT'S FIRST-MOVER ADVANTAGE

Importance of Choice of Equipment Manufacturer

Business strategies such as customer orientation and continuing R&D must tie in with *operational strategies* to achieve outstanding performance. This is especially true in the upper-end industries of semiconductor-related manufacturers. Since semiconductor manufacturing equipment is very expensive, if the wrong equipment is bought, the company must deal with costly disposal costs, waste materials and delays in the production process. Once a chip-maker decides to use equipment from a certain provider, the process from document delivery, shipping, installation, fine-tuning and testing, to starting production is time- and labour-consuming, making costs extensive and hard to estimate for clients deciding to change equipment. Semiconductor manufacturers therefore always make a careful evaluation before choosing their equipment provider: once the decision is made, the relationship becomes ingrained and difficult to change. The first to open the door to the semiconductor manufacturing equipment market is therefore the most likely to gain a lasting position and receive continuing orders.

Competitive Strategy

AMT was one of the first equipment suppliers to expand in the Taiwan market, and thus scored an immediate major victory. To maintain its advantage, employees then made regular and extensive efforts to secure customer orders. With good R&D technology and complete service networking, AMT endeavours to meet customer needs, making customers willing to rely on AMT's technical support and service, and continuing to place orders. This forceful competitive strategy is especially important to Applied Materials, who continuously research and develop new products. Customers usually hesitate to buy a new product when it is first introduced to the market, since they are dubious about the stability of the machines. AMT attempt to introduce new products to customers through aggressive sales, supplemented with complete technological service and R&D capability – on-site installation and testing, around-the-clock support, comprehensive customer service, a continuing improvement plan, a cooperative R&D project, a total component management project and an engineering and technological training centre. These services are intended to resolve all possible problems the new equipment may have, and help it to achieve expected throughput and get up to speed in the shortest time (Yen and Liu, 1996). Once machine quality is stable, AMT starts to produce and sell it on a large scale. The principle of economies of scale means that production costs decrease as more machines are produced, so AMT can pass on lower costs to its customers and devote more capital to developing a new generation of equipment. This becomes a self-perpetuating cycle, and AMT can gain two advantages at once.

Obtaining first-mover advantage, the spirit of customer orientation, and continued R&D form the three pillars of the company's strategy, and has left competitors lagging far behind. Sales teams are like marines who take over a beachhead to be supported later by sophisticated weapons and equipment (R&D) and an army of rear-service units (customer support teams). This has helped AMT establish and consolidate its position as the leader of Taiwan's semiconductor manufacturing equipment providers.

WINNING THE HEARTS OF EMPLOYEES

Why is Efficient Use of Human Capital so Important?

Efficient use of human capital is a critical element for AMT as it works towards its 'total solution' through being close to customers, developing advanced semiconductor production ability and achieving world-class team performance. For most companies, especially hi-tech companies in HSIP, the quality (good and bad) of HR is closely related to both HR development and the company's compensation and benefit systems. In HSIP, where hi-tech talent is always in great demand, recruiting from other companies is relatively common, and workers often change jobs. If a company does not manage its people well and provide attractive compensation and benefits and promotion opportunities, it will be difficult to recruit and retain workers. In such a competitive industry, a company cannot survive without a top-tier workforce and efficient human resources management.

Employment Philosophy

AMT considers its HR as its most important capital. 'To respect, take care of, and cultivate workers' is AMT's basic philosophy in its human resources management (HRM) system (author's own data). The company strives for foreign firm-level efficiency, Chinese-style human-oriented management, good compensation and benefits, maintaining a good learning environment, respect for individual growth and profit-sharing with employees (Chen and Keng, 2001).

Employee Training Plan

AMT has adopted an 'employee training plan' as the first component of its HR development policy. Each employee receives on average at least 40 hours of training, and has the opportunity to go abroad for training courses every six months. This policy can satisfy employee needs for learning and growth, and help them form a broader world-view. Training courses are also closely connected to employee promotion and development. Employees must complete certain training courses at their current job level to qualify for job promotion. All employees must receive training courses on such topics as new employee orientation, corporate morality, and quality control program. Required training courses for 32nd- and 33rd-level employees are interpersonal relationship and communication and customer service skills. Communication and teamwork is required for employees at the 34th level, while managers at the 35th level or above need to take courses on interviewing skills, the basic concept of management, performance management, situational leadership, financial management and strategic strategic. After receiving all the required training, the employee is qualified for promotion.

Talent Development

AMT established its model for talent development based on the values and characteristics of successful managers, evaluation of potential and specific targeted skills. This model contains ten groups of abilities (see Table 7.9).

Table 7.9 AMT's talent development model

- Change management
- Cognitive thinking/problem solving
- Communication
- Customer orientation
- Global operation perspective
- HR management
- Leadership
- Personal efficiency
- Project management
- Value building

Source: Chen and Keng (2001).

According to this model, the HRM department cooperates with managers of other departments as well as senior workers to develop a training framework to improve the quality of AMT's human capital.

Employee Career Planning

Based on the talent model, AMT in 1998 proposed the 'individual development plan' (IDP). Each employee must make a self-evaluation according to his/her own special skills and career interests. Supervisors also evaluate and suggest abilities and skills needed by their subordinate. Then the supervisor and subordinate discuss and develop an IDP according to both evaluations, and choose three out of the ten groups of abilities in Table 7.9 as the employee's annual goal of development. The outcomes are evaluated annually at the third quarterly meeting and the evaluation results become part of the employee's educational training outcome appraisal for that year. Supported by high-level management and ardently participated in by employees, the success rate of this plan is over 90 per cent, and the effects on the business are significant.

Appraising Training Performance

Through questionnaire investigations, self-reports, paper-based tests, learning quizzes, group evaluations, briefs/workshops, annual appraisals, and follow-up learning, all training is holistically evaluated. These measures are adopted to improve employees' work effectiveness, rather than to waste their time and business resources.

Table 7.10 AMT's government awards for HRM

1997	Human Resources Development and Training Efforts Award
1999	Excellent Labour Training Award
2000	Excellent Employee Relationship Award
2001	Excellent Performance in Occupational Safety and Health Award

Source: Chen and Keng (2001).

Compensation and Benefits Schemes

To motivate employees, AMT experimented with several kinds of compensation and benefit schemes. Some measures included introducing employee stock options, three-year interest-free auto loans (up to NT$360 000), group insurance (in addition to labour and health insurance, employees and their dependents also enjoy free group insurance), more holidays and vacations (employees are given up to 30 days of holiday and vacation per year, exceeding labour law requirements), meal premiums (free breakfast, half-price lunches and dinner allowances), allowances for special occasions (birth and marriage allowances, important holiday allowances and birthday cash gifts), crèches for breast-feeding female workers, temporary employee dormitories, annual domestic/overseas travel allowances for employees and dependants, a mental and physical health plan (including employee health examinations, on-site doctor diagnosis, professional contract counsellors and a well-equipped fitness centre with professional coaches).

These compensation and benefit schemes are designed around specific employee needs: 80 per cent of engineers want to have a car straight after graduating from school, so AMT began to offer three-year interest-free auto loans; since many employees do not have time for breakfast, AMT provides a free Chinese and Western meal. To improve employees' mental and physical health, AMT proposed 'green hill' plans to encourage employees to exercise, have quality time with their family and join employee get-togethers so that employees can achieve a balance between work and family while they develop their own careers. AMT's dynamic HR development system and compensation and benefit measures have received a number of awards from the Taiwan government (see Table 7.10).

Results of a survey conducted by *Management Magazine* (May 2001) showed that college graduates ranked AMT sixth among Taiwan's 'Most Preferred' companies to work for. The quality of AMT's human resources is good, with 47 per cent of the workforce holding postgraduate or doctorate degrees, and 88 per cent undergraduate degrees, demonstrating that AMT's management is efficient in keeping talented people working for the company – another competitive advantage for AMT.

Building a Local Image

Fitting in with local culture and industry is the most important way for foreign companies to be accepted by local businesses, the academic community and government and, in turn, to expand its business. AMT has made great efforts since 1989 to promote its business localisation, including measures to boost cooperation between industrial–academic communities, cultivate local providers, promote environmental protection and improve public welfare.

Cooperation with Academe

To enhance cooperation with the academic community, AMT provides universities and colleges with numerous resources, and has earned great praise for its efforts. In 1998, AMT and the National Chiao-Tung University (NCTU) began a one-year cooperation project, in which AMT's training centre invited two professors from NCTU to receive a series of complete semiconductor production training courses, and allowed them to take what they had learned from the courses back to the university to offer related classes to students. AMT also provides NCTU students with 300-hour practical training opportunities. Students can go to the training centre to practise machine operation, disassembly and maintenance in order to cultivate their abilities and experience in repairs and equipment R&D. In 1998 AMT also set up 'scholarships for technical elite cultivation' at six universities: NTHU (National Tsing Hua University), National Taiwan University (NTU), NCTU, NCKU (National Cheng Kung University), CYCU (Chung Yuan Christian University) and NTUST (National Taiwan University of Science and Technology) in Taiwan. It also offers training courses in the semiconductor production process, and equipment, to scholarship winners. In 1995, following a plan by Chiam Wu, AMT donated a set of high-level semiconductor equipment to NTHU in order to provide faculty and students with the practical experience of operating the machines. In 1999 AMT also contributed the series of P5000 semiconductor facilities and NT$10 million to NTHU to assist in the establishment of a semiconductor laboratory. These actions were integral for the cultivation of local talent in semiconductor equipment.

Local Support Base

Since 1994, AMT has actively promoted a project to develop a local supplier base for critical components, to help develop home-grown precision machine factories and component suppliers providing critical parts and raw materials for semiconductor equipment. This project helped semiconductor equipment manufacturers to gain cheaper components faster and also helped the domestic precision machine industry to upgrade. In the 2000 'flowing water' project, AMT signed a 'strategic alliance in the supply chain of semiconductor production and critical components' contract with local component providers it had helped to cultivate, including Gongin, Annex, TOPCO, Eternal and Rongzhong, to establish a complete local supply chain for Taiwan's semiconductor industry. This project also improved the industrial distribution of critical components for Taiwan's semiconductor equipment manufacturers, boosting international competitiveness.

AMT was additionally successful in integrating environmental protection with its business operational strategies (see Table 7.11). These efforts have won a number of awards (see Table 7.12).

AMT also promotes many kinds of public welfare events, including a charity concert raising funds for the 9/21 earthquake (21 September 1999)

Table 7.11 Environmental protection at AMT

- Recycling and reusing waste water
- Recycling packaging materials from work sites and storage
- Reducing the use and concentration of chemicals and solvents
- Decreasing pollutant discharges
- Tracking, cleaning and shipping of poisonous substances
- Constructing environmental monitoring facilities
- Holding events to advocate environmental protection on Earth Day

Source: AMT (2001).

Table 7.12 AMT's environmental protection awards

- Water Conservation Award (Economic Department)
- Industrial Excellence Award (Economic Department)
- Industrial Waste Minimisation Outstanding Award (Economic Department)
- Corporate Environmental Protection Award (Environmental Protection Department)

Source: AMT (2001).

in Taiwan and an emergency medical first-aid training plan, the brainchild of the HSIP administration. It also donates money to support charitable institutions in the Hsinchu area (such as the Chinese Fund for Children and Families/Taiwan and Mental Retardation Centre), and holds art and literature workshops. These events helped AMT to receive awards from HSIP and the Council of Culture Construction, and to enjoy a sparkling public image for its social welfare policies.

THE TRIANGULAR RELATIONSHIP BETWEEN TAIWAN, MAINLAND CHINA AND AMERICA: AMT'S FUTURE CHALLENGES

AMT has been running smoothly for over a decade. Advantages such as solid support from its parent company, an opportune entry to the Taiwan market, Chiam Wu's unique leadership style, a customer-orientated philosophy, continuing R&D, first-mover advantage, good HR development and compensation and benefit systems, and continuing promotion of business localisation and a good social public welfare image all helped AMT to make breakthroughs and improve profits continually. Yet since 2001, changes in both the ecology of the Asian semiconductor industry and the high-level management of AMT have altered the landscape and diluted some of the factors that were critical to its success. What will AMT's future be?

Market Growth

In 2001–02, the growth rate of the global semiconductor market was −13.5 per cent. The Taiwan semiconductor industry was also affected by the decrease in chip demand: ITRI figures show that, compared to 2000, the output value of the Taiwan IC industry in 2001 decreased 26.6 per cent. Though experts predicted that starting in 2002 the global semiconductor industry would revive, Morris Chang also pointed out that the global semiconductor industry has historically had a five-year cycle of prosperity and decay. Market decay started from the second quarter of 2001, and is currently on its way to recovery. Recovery of the global semiconductor industry does not necessarily lead to an equivalent revival of the Taiwan semiconductor industry, however, given that the industry in China is starting to grow and threaten Taiwan. Even though the global semiconductor industry was weak in 2001, the annual IC throughput of the Asian area for the first time exceeded that of the United States (the former accounting for 28.4 per cent, and the latter 27.1 per cent), and became the world's largest

IC provider. This was mostly the result of the upsurge of investment in the Chinese semiconductor industry, which in turn led to the rise in semiconductor output and of Asian market share.

In China, many factors, such as cheap land and low-cost water and electricity, abundant manpower (especially low-level manpower) and strong government support (establishment of new industrial parks, large loan provisions, tax incentives) attracted US-based AMD and IBM, Japan-based Toshiba and Singapore-based Chartered (Chartered Semiconductor Manufacturing) to invest. A number of Taiwan entrepreneurs also quietly began semiconductor businesses in China. Unlike the slowdown in the Taiwan semiconductor industry in 2001, the Chinese semiconductor industry boomed, with a growth rate close to 40 per cent. Merrill Lynch estimated that the average annual growth rate of the China semiconductor industry would be about 25 per cent over 2002–06, helping it absorb a market share of US$43 billion. US experts predict that China will become the world's second largest IC market by 2010, following only the United States.

Falling Demand

Demand for semiconductor manufacturing equipment always rises and falls in line with the business cycle of the semiconductor industry. With the recession in the semiconductor industry, the demand for production equipment decreased from US$47.7 billion in 2000 to US$31.5 billion in 2001, a 35 per cent decrease. In Taiwan, the decrease reached 55 per cent, far exceeding the global average. AMT's total contribution to sales of global Applied Materials decreased from 20 per cent in 2000 to 15 per cent in 2001. China was the only area showing a positive growth in semiconductor equipment demand around the world in 2001–02, and the development potential for semiconductor equipment in the China market is unlimited. James Morgan, president of global Applied Materials, said that total purchases of semiconductor equipment in the China market will reach US$500 million, and the growth rate of sales revenues will be about 500 per cent. China will be the next market to demand a great quantity of semiconductor manufacturing equipment and the China market has become an important place for Morgan to develop his business.

Power Structure

Anyone who can dominate the Taiwan and China semiconductor markets will win the role of global Applied Materials' chief Asian operator. Wu foresaw the potential of the China market and the limitations of the Taiwan market, and made efforts to persuade the parent company to fight for market share in China. Wu believes that AMT, rather than Applied Materials China (AMC), should have the right to handle Taiwan chip-makers' equipment purchases when they establish their IC companies in China. However, David N.K. Wang, vice-CEO of global Applied Materials and president of AMT, disagrees with Chiam's strategy of westward expansion, and took the side of AMC's general manager, Nai-xin Lin: Wang put a lot of effort into helping AMC expand in the China market and did not intend to allow Wu to contest it without a fight.

This conflict over the China strategy has been fierce, and Wang seems to have come out on top. Relying on his good interpersonal relationships with government officers in Taiwan and China, Wang was able to pull strings with James Morgan and with the Taiwan and China governments. Wang also obtained orders from chip-makers such as Semiconductor Manufacturing International Corporation (SMIC) and Grace in the Shanghai area. In October 2000, David Tu was assigned to be AMT's general manager, and Applied Materials held a grand ceremony to celebrate the establishment of its Shanghai office, voicing its determination to develop its business in China.

David Tu graduated from National Cheng Kong University, majoring in Material Science. He received his master's degree in industrial management and a doctoral degree of material science from New York State University, and became a consultant of Academia Sinica. In 1994 he founded Duratek Inc., where he continues in the post of president and general manager. The original business focus of Duratek was to research and develop artificial joints, but it shifted to producing semiconductor manufacturing equipment after three years. Tu proved to have a solid technical background and was able to predict market trends precisely; he was noticed by David Wang and became a member of the Wang camp. Chiam Wu was promoted to be AMT's vice president to assist Applied Materials in strategic planning for the Asian area and she retained her position as vice president of the global Applied Materials group. On the surface, David Wang touted Chiam's ability:

> I am glad that Chiam has taken on further responsibilities, and is helping too in strategic planning for the Asian area. Chiam has been responsible for establishing stable and close partnerships with Taiwan clients. For the past decade, Chiam has led AMT to grow rapidly and helped Applied Materials become the leading contributor to the Taiwan semiconductor industry. (AMT, 2001c)

In fact Chiam's promotion was actually a demotion and led to the loss of her predominant power in AMT. By means of his agile political manoeuvring, good interpersonal relationships with cross-Strait high-level government officers and his efforts running business in the China market,

Wang won the so-called 'cross-Strait inter-subsidiary battle' and gained the operational territory of the greater Chinese area (Taiwan and China).

OTHER CHALLENGES

Though re-engineering has temporarily come to an end, new challenges are appearing. With China's participation in the semiconductor market, there is bound to be a change in the ecology of the Asian semiconductor industry. The Taiwan semiconductor industry still has prospects, but the China market has become more attractive. How can Wang court both the Taiwan and China market? What role should he play when facing the increasingly strong competition between AMT and AMC? Should AMT or AMC get the business if the Taiwan government allows investment in 8-inch wafer production in China?

The redeployment of high-level managers has brought AMT a future full of uncertainty. For a decade, Chiam Wu had been a prominent leader who ran AMT impressively and won the support of employees. Wu's good relationship with customers also brought AMT numerous orders. David Tu was formerly the president of Duratek, a small-scale company with fewer than ten employees and which is eyeing NT$1.9 billion for its 2005 annual revenue. Compared to Duratek, AMT had 900 employees and over NT$60 billion revenue in 2000. Such a difference between Tu's former and current companies will bring him some serious challenges and he has to start establishing relationships with customers from the beginning. How is he going to obtain support from his subordinates and customers in order to lead AMT to another financial peak? When Applied Materials announced it intended to lay off 2000 employees in 2001, AMT changed the status of over 90 per cent of its administrative personnel and about 20 per cent of its engineers to contract workers. If Tu is going to start to downsize AMT's personnel, will employees be willing to work for AMT and continue to live up to the company's spirit of 'customer orientation' and 'continuing R&D'? How will Tu react to being sandwiched between Chiam Wu and David Wang? The answers to the above questions will be of critical significance once Tu has taken up his position.

REFERENCES

Addison, C. (2001), *Silicon Shield: Taiwan's Protection Against Chinese Attack*, Dallas, Texas: Fusion Press.
Applied Materials Taiwan (1998), *Corporate and Product Information*.

Applied Materials, Inc. (2000), *Corporate and Product Information.*
Applied Materials Taiwan (2001a), *About Applied Materials*, www.amat.com.tw/about/index.asp
Applied Materials Taiwan (2001b), *Customer Support and Service*, www.amat.com/tw/about/index.asp
Applied Materials Taiwan (2001c), *News Centre*, www.amat.com.tw/about/index.asp, October 18.
Applied Materials Taiwan (2001d), *Production Centre*, www.amat.com.tw/about/index.asp
Chen, Z.H. and Keng, C.I. (2001), 'The employee training system in AMT', in T. Lee (ed.), *Human Capital Management in High-Tech Industry*, Taipei: CommonWealth Publishing.
Chuang, S.Y. (2001), 'The interview of AMT's CEO: Ms Wu', *Global Views Magazine*, February, 34–48.
Yen, C.F. and Liu, S.C. (1996), 'The mission of AMT: Customer orientation', in C.Y. Li (ed.), *The Successful Transformation of Medium- and Small-sized Business in Taiwan*, Taipei: Chinese Management Association.

8. Philips Semiconductors Kaohsiung (PSK)

Chia-wu Lin

THE PAST AND PRESENT OF PSK

For most Taiwanese people, the word 'Philips' stands for an internationally famous brand of light bulbs and home appliances. Many people also know that Philips has engaged in trade in Taiwan for a long time, establishing a manufacturing factory and an R&D centre. But few people are aware of the place of Philips Taiwan Semiconductors Kaohsiung in the Philips Taiwan Enterprise system. Philips Taiwan Semiconductors Kaohsiung – or, as it is called by its in-house staff, Philips Semiconductors Kaohsiung (PSK) – was the earliest subsidiary company set up by Philips in Taiwan. PSK was established in 1966, located in the processing zone in Nan-tz Kaohsiung. Originally there were two factories, one for passive components, the other for integrated circuits (IC). The integrated circuit factory, the parent company's important IC headquarters and the IC automatic technology centre, was set up in 1966, and engaged in advanced package assembly and IC testing. The annual production of ICs in 1995 was close to 800 million pieces, and PSK plays the role of production headquarters in the Philips Electrical Components business department, one of Philips' three most important centres of packaging and testing of wafers.

THE PASSIVE COMPONENT PLANT

Set up in 1967, the passive component plant initially produced Magnetic Circle Pads, and subsequently various kinds of passive component products were imported, including Multi-Layer Chip Capacitor (MLCC) which made the plant the only manufacturer which could operate consistently from upstream, the synthesis of powder, to downstream, completion. After the establishment of the Asia Pacific Business Centre, the passive component plant became the parent company's production-marketing and fine ceramics electronic component development centre. The main products were Surface-Mount Device (SMD) electronic capacity components; annual productivity was 20 billion pieces, mainly supplying the electronic components market in the Asia Pacific region.

CHANGES IN 2000

In 2000 this situation changed. Philips sold the passive component plant to Yageo Corporation, renamed Phycomp, but still specialised in R&D and production. Despite the absence of a passive component department, PSK Ltd's performance was still outstanding, being listed in the top ten Taiwan manufacturing industries (fourth in 1999, second in 2000). By 2000, PSK's annual production had expanded to 1.4 billion pieces of passive components and its overall sales attained some US$88 billion. PSK had become Philips' worldwide semiconductor IC packaging/testing plant, with the highest sales and the leading technologies. PSK was also Philips' logistics headquarter in the Asia Pacific region engaged in ordering, sales, logistics and warehousing.

ATO

Although PSK had outstanding achievements, its subsidiaries did not have great autonomy. As one of its in-house mid-level supervisors said jokingly, PSK was merely an OEM of Philips' semiconductor business department, mainly engaged in the middle or the final processes of passive component production, IC packaging/testing. In the organisation chart of the semiconductor business department (see Figure 8.1), it was called ATO (the Assembly, Testing Organisation). There were two other ATOs in Philips' global domain, located in Bangkok, Thailand (PST) and in Manila, Philippines (PSC).

PSK in Philips' global tactics played a packaging/testing plant role, which was included in the six global product divisions (consumer electronics, home appliances, luminaries, semiconductors, electronic components and medical systems).

ORGANISATION AND STRUCTURAL POWER ORIENTATION

PSK and Philips Taiwan

In 1966, under the coordination of Mr K.T. Li, the senior advisor to the Office of the former President of Taiwan, an Export Processing Zone

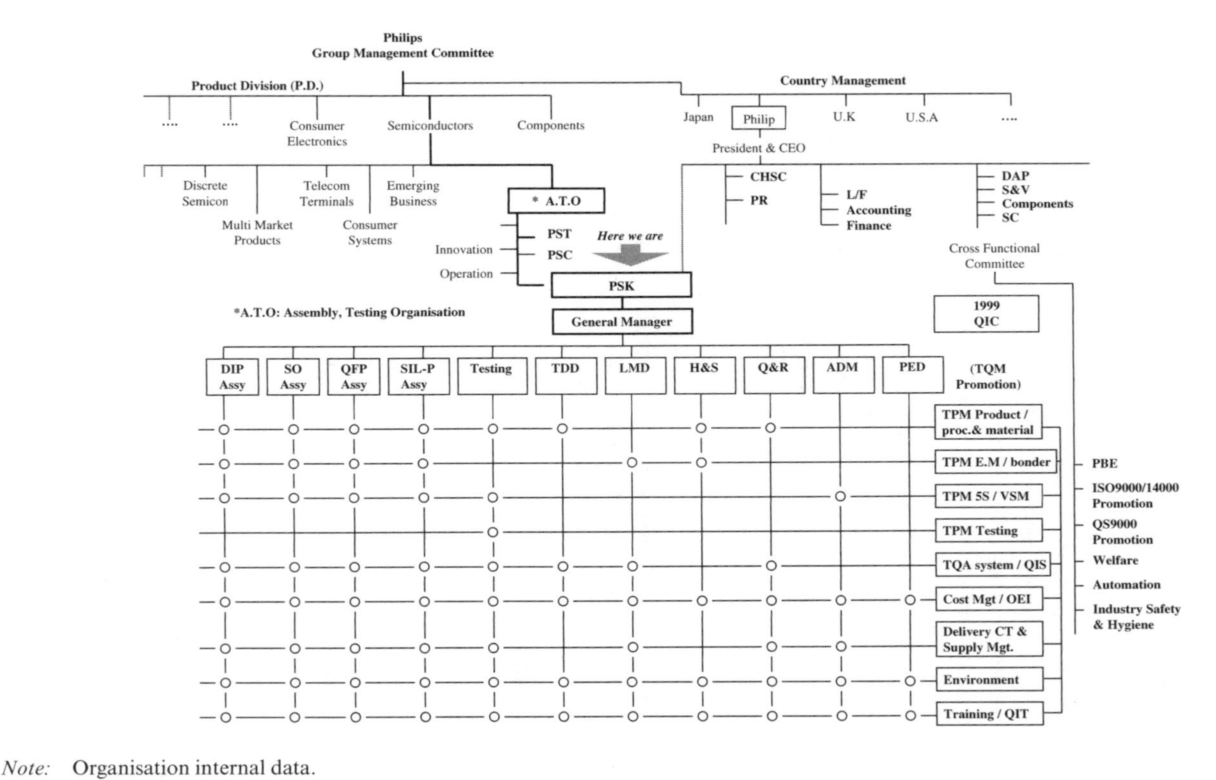

Note: Organisation internal data.

Figure 8.1 PSK organisation position

(EPZ) was set up in Kaohsiung. In the following 30 years, Philips NV frequently invested in Taiwan, establishing Philips Taiwan Co. Ltd, Philips Electronic Building Elements Industries Taiwan Ltd and Philips Taiwan Electronic Industry Co. Ltd. At its peak in 1997, Philips had seven plants and 12 600 employees. In the 1990s there was an industry upgrade and transformation: with continuing inputs of skill and capital, Philips Taiwan became the largest foreign invested company in Taiwan, with sales of NT$204.3 billion in 2000.

Philips Taiwan produced mainly consumer electronic products and key components, including audio and video products, lighting, home appliances, consumer communication (cellular phones), monitors, computer peripherals, electronic components, semiconductors, commercial electronics products and medical appliances. In the mid-1990s there were seven plants in Chungli, Taoyuan, Hsinchu and Kaohsiung. Products included monitors, lamps, cathode ray tubes, integrated circuits and passive components. In response to the strategic changes of moving into Mainland China, however, the seven plants were reduced to three plants located in Dayuan (lamps), Chungli (monitors) and Kaoshiung (semiconductors) in 1999. Philips Taiwan is still the important Asia Pacific and global business centre of Philips worldwide. The headquarters set up in the Taiwan business centre were the headquarters for semiconductor Asia Pacific, Electronic component Asia Pacific and Global Monitor. In 2000, Philips NV set up its first Asia Pacific R&D Lab in Taiwan, its sixth Global R&D laboratory. Figure 8.2 shows Philips' sales in 1985–2000.

Power Orientation

In the current organisation structure, PSK is mainly under the direction and control of the Semiconductor business department. Philips Taiwan Co. Ltd, located in Taipei, plays the role of the 'landlord' in charge of the businesses and servicing of administration, tax and decree laws. It looks like a matrix organisation model but actually it is a product-divisionalised model, but one completely different from that only 5–10 years ago (please see Figure 8.5 on the PSK organisation structure). From 1986 to 1996, Philips Taiwan was under the direction of the president, Mr Y.C. Lo, who is Taiwanese. Lo joined PSK as an IC engineer. After holding the positions of general manager of PSK, general manager and vice-president of Philips Electronic Building Elements Industries Taiwan Ltd, he was promoted to join the GMP-Group Management Committee at the policy decision centre in the Netherlands' head office in 1996. The next two presidents of Philips Taiwan after Lo, Gerard Kleisterlee and P. Zeven, carried out the strategic plans of the parent company precisely, transforming the Philips Taiwan

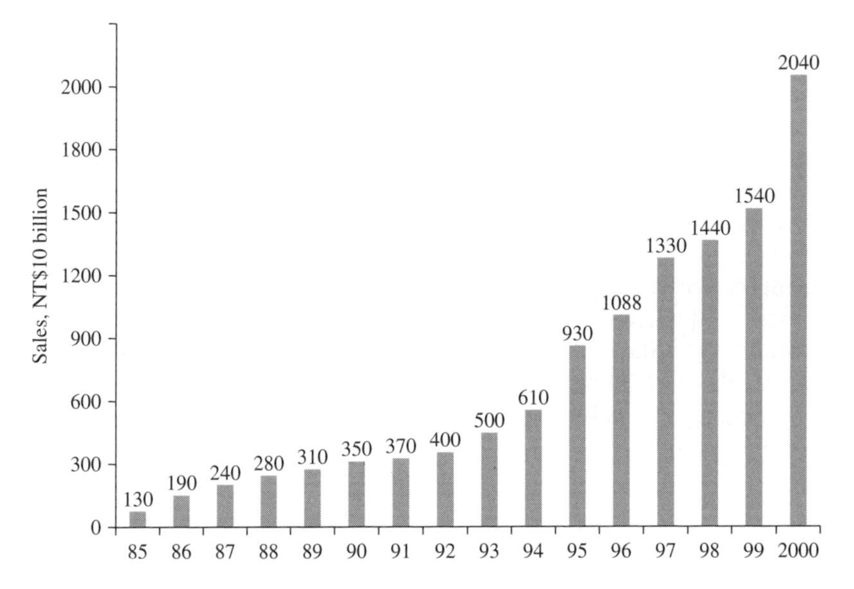

Figure 8.2 PSK sales, 1985–2000

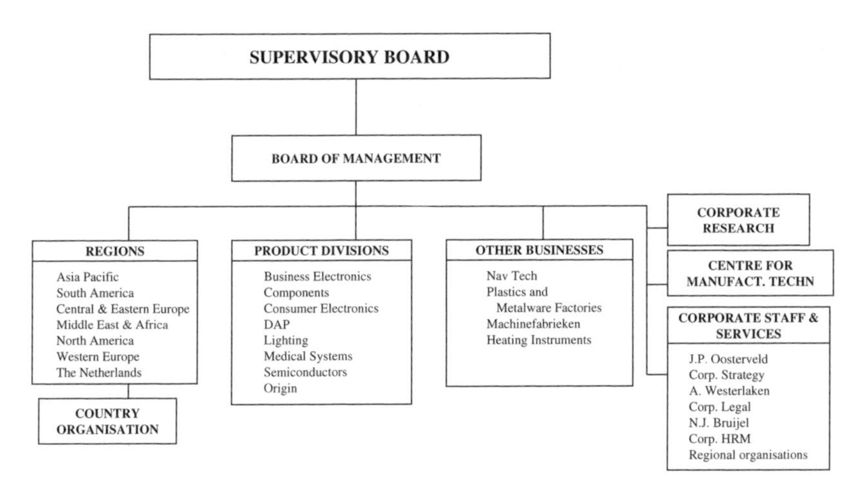

Figure 8.3 Philips organisation chart

headquarters in Taipei into an office for administrative services. The organisation structure of Philips worldwide has a tendency to follow the product division organisation model (see Figure 8.3) so that Philips Taiwan is relegated to the management of Philips in Mainland China.

The headquarters of ATO was never transferred to Taiwan, and is located in Singapore. This aroused strong feelings at PSK and Philips NV tried to redress the balance of power (ATO had three plants); they possibly also felt that the infrastructure of global logistics in Taiwan was not as good as that in Singapore. Now Philips is finally entering the market of Mainland China, and although PSK is essential for Philips Semiconductor Division's packaging/testing, Philips still insisted that the Semiconductor Division management centre be set up in Singapore. PSK's high-level talents in technology are still needed in the Asia Pacific region for such things as the establishment of new plants and training education in Singapore and the Philippines. PSK also played the role of 'technological parent plant' of Philips Semiconductor Division for packaging/testing, providing high added-value and a high level of technology and technology transfer capability.

PHILIPS TAIWAN AND THE DEVELOPMENT OF PRODUCTION

Wafer Foundries

PSK passes most production to TSMC for wafer foundry producing (see Chapter 5) and then redelivers it to Philips Taiwan for final packaging/testing processes. TSMC is the leader of the foundry industry and Philips NV has a share of some 20 per cent in TSMC. Philips and TSMC are currently in partnership to build a foundry plant in Singapore.

The business department of PSK focuses mostly on distribution in Taiwan and part of the R&D of consumer IC is placed there. The packaging/testing plants are exclusively for PSK. Even when outside competitive packaging/testing companies can provide lower prices, Philips still have their products manufactured by their in-house factory for the sake of quality and strategy: PSK, PST and PSC are called the 'in-house factory' of PSK. Figure 8.4 shows the role Philips Taiwan plays in the strategy and process of the parent company.

SEMICONDUCTOR MANUFACTURING TECHNOLOGY

Philips Taiwan has constantly introduced and transferred advanced semiconductor manufacturing technology to Taiwan. Philips also set up a department of technology development, significantly aiding the ability to

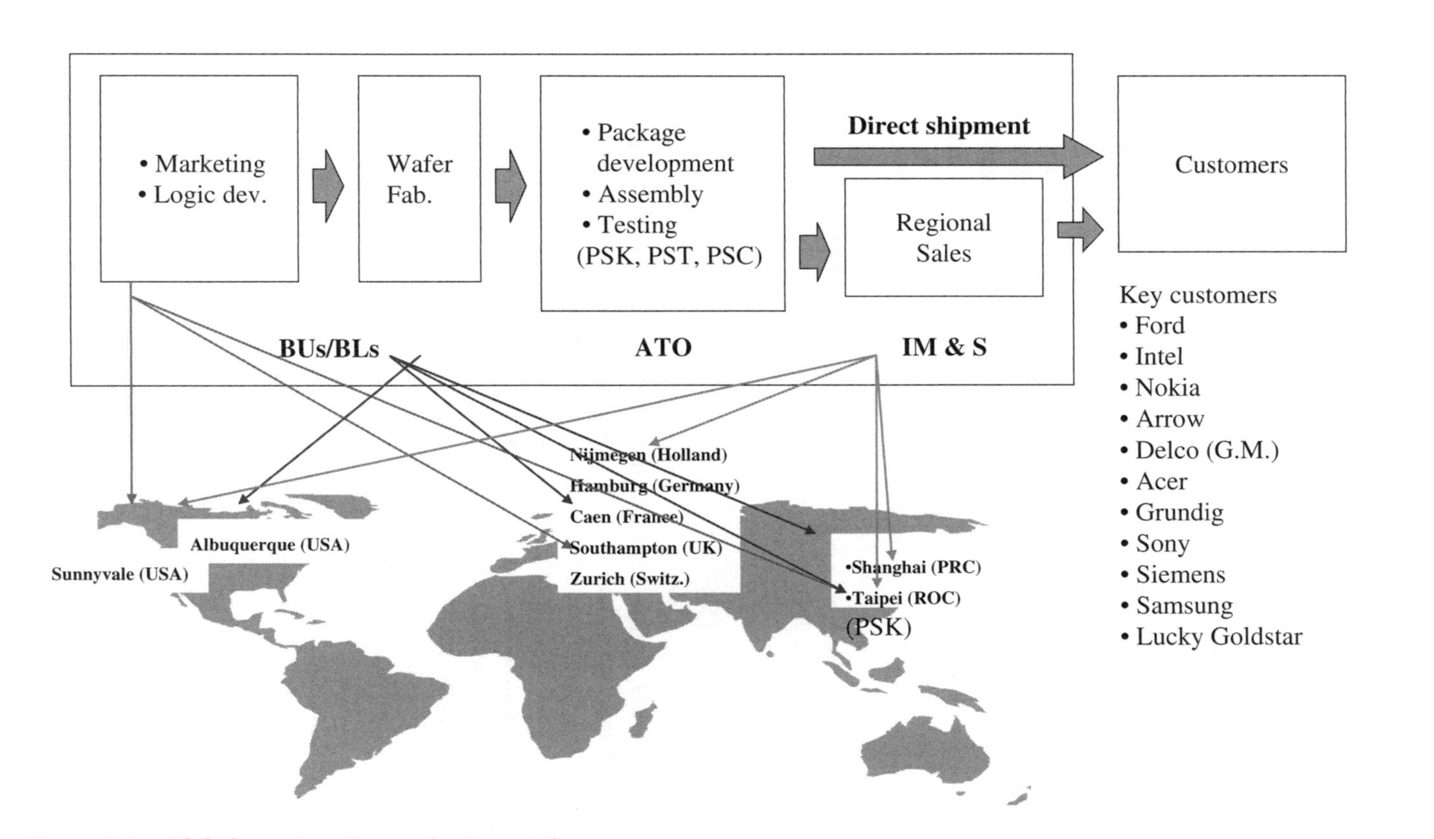

Figure 8.4 Philip's semiconductors business infrastructure

manufacture new products and technology such as Ball Grid Array (BGA), Multi-Chip Packaging (MCP), Die-to-Die package technology and Quad Flat Package (QFP). In its drive to achieve the leading position in quality of production, cost and delivery time in the industry, Philips also cooperates with ITRI (see Chapter 2) and Taiwanese universities to develop new techniques of manufacturing.

Only in the mid-1980s did Philips Taiwan depart from the labour-intensive activities which had characterised its early stage of development (1966–75). This saw the company evolve from simple assembly (1960s) to a television industry (1970s). In the 1980s Philips Taiwan coordinated with the Republic of China (ROC) government in major industry development and then shifted its focus to the development of an electronic information industry. However, in the mid-1990s, following the strategic direction of head office, Philips Taiwan began to place more emphasis on the development of a consumer electronics industry alongside the electronics information industry. From 1970, Philips Taiwan has been closely involved in increasing the competitiveness of Taiwan's electronic industry: in 1988 its output of resistors, CRTs and other components was the largest in Taiwan.

MULTIMEDIA INFORMATION PRODUCTS

From the 1980s, Philips Taiwan gradually upgraded its production from 'capital/technology-intensive' to 'intelligence-intensive', with a strategy of transforming itself from a home appliances manufacturer to a multimedia information product supplier. Major business included IC designing, packaging and testing and manufacturing monitors and their CRTs. The company later invested in the Taiwan Integrated Circuit Company and started manufacturing Notebook PCs (see Table 8.1).

Table 8.1 Philips Taiwan: IC development, 1969–74

Year	Product	Company
1969	Packaging	Philips Taiwan semi-conductor plant
1974	Packaging and testing	
1989	Chip testing, packaging and testing	
1986	Investment in producing chips	Joint Venture Incorporated: TSMC
1986	Product design	Taipei head office semi-conductor design centre

STRATEGIC ORIENTATION OF PHILIPS TAIWAN

PSK was the oldest plant set up by Philips in Taiwan, the packaging/testing plant with the largest capacity and effectiveness, and the main packaging base of the Philips Taiwan semiconductor department, doing packaging and testing for high-level, complicated or other low-quantity semiconductor products (see Figure 8.5). The low-level semiconductors are manufactured by the PST and PSC plants in the Philippines or Thailand. Why was PSK positioned for the packaging/testing stage in the manufacture of semiconductors? This is related to the direction of industrial development in Taiwan in recent years: cheap labour and low efficiency in the early period, manpower of guaranteed quality in recent years and the 'cluster effect' of the Hsinchu Science-based Industrial Park (HSIP, see Chapter 3). Many electronic companies also made Philips NV invest NT$16 billion on plant and equipment to expand yield in PSK, even when PSK was facing a 'China fever' environment. PSK has thus become the semiconductor packaging/testing plant with the most sales and the most advanced technology, manufacturing 1.4 billion semiconductors per year and having total capital amounting to more than US$88 billion.

In the light of Philips' gradual bowing out of photoelectric components (such as the winding down of the Chupei and Dapon plants), why does Philips Taiwan retain the favour of the parent company and continue to attract investment? This is chiefly related to the characteristics of the semiconductor industry. Techniques and yield have great influence on the quality and the cost of products. But the ratio is low when it comes to the costs caused by locating near to the market (decreasing transportation costs, for example). For photoelectric components, however, the ratio of cost caused by locating near to the market is very high, so it is impossible to supply monitors made in Taiwan only in the Taiwan market. The PC market in Mainland China is developing rapidly, making a move to Mainland China unavoidable.

Even though PSK plays an important role in Philips' semiconductor department, its cultural and regional influence is not that important, for this is what high-level directors from PSK call a 'closed system market'. As long as the cost of labour in Mainland China is still cheap, and the quality of manpower in high-tech is rising, the enormous market is a deadly attraction. PSK has set up a packaging/testing plant in Suzhou in Mainland China and is bowing out of its 'Semiconductor research centre', subordinating it to the semiconductor department. Several of the research members from this centre are now gradually being sent to Mainland China as the centre winds down.

From PSK's global viewpoint, its chief goal is still the 'specific' market, but it is impossible to invest money in R&D on technology and

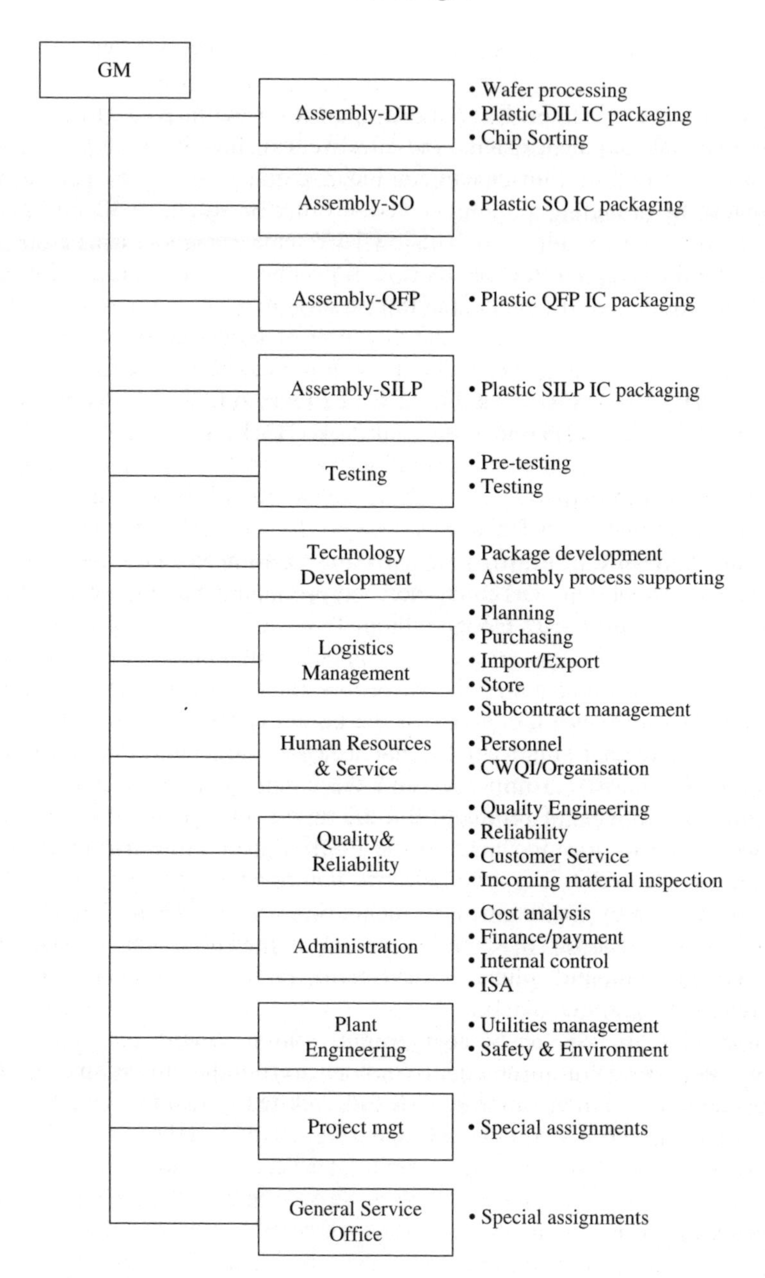

Note: GM is General Management.

Figure 8.5 PSK organisation chart, 1999

manufacturing simultaneously. PSK has therefore chosen a different strategy from that of Intel and AMD. Its chief rival is Siemens, but Philips Taiwan has a strong position in communication products and IC design. When competing against Japanese companies such as Sony, PSK products supply not only the internal needs but can compete to get orders from outside, by providing a complete system solution to clients. But it might buffer different subsidiaries of other departments within the same group by providing a complete system solution. So deliberate integration is a necessity across all departments in the Philips group.

PHILIPS TAIWAN'S COMPETITIVE ADVANTAGE: BELIEF IN ADHERENCE TO QUALITY

A 'Culture of Quality'

Philips Taiwan was awarded the Japanese Deming Prize in 1991 and the Japanese N-prize in 1997. The former president, Mr Y.C. Lo, a local Taiwanese professional manager from PSK, has always appreciated the value of awards for quality, instilling a belief in the pursuit of quality and constant improvement in PSK managers. The concept of 'quality' is a shared language, challenge, way of working, logic and belief exclusive to Philips Taiwan: although PST or PSC can keep pace with Philips Taiwan technically, they have not developed this unique culture of quality. The culture does not change with different leaders: presidents may come and go, and processes may be different, but enthusiasm for pursuing quality remains.

In 2000, the slogan of the 'quality month' in Philips Taiwan was 'Intuitive knowledge, good behaviour and conscience'. This slogan symbolises the company's goals and determination to raise competitiveness in the business and to make sure their competitive advantage is retained. The emphasis is on a spirit of comprehensive improvement of quality and continuing R&D of automatic technology and equipment. In 1998 Philips Taiwan was awarded an Achievements of TDP award by the Taiwan government.

QUALITY CIRCLES

After introducing quality circle (QC) activities in 1978, the company set up 17 QICs (Quality Improvement Committees) and 141 QITs (Quality Improvement Teams) to introduce the activity of improvement for all members, planting the idea of all-round quality deeply in all members' minds. In December 2000, Philips Taiwan tried to promote the improvement of quality on the production line, with their so-called 'Ding-Dang Circle'. This quality circle won the National TQM Silver Medal in 2001 as a result of their excellent performance. In the process of improving the Ding-Dang Circle, they not only convene members involved but invite equipment and block engineers to form a cross-department team in order to strengthen improvement. With teamwork from all departments, dissidence among departments is ended with the aim of achieving excellence and quality.

THE FUTURE

Philips Taiwan is reducing its scale in Taiwan. But in 2000 production was three times that of 1997, even when nearly 5000 employees had been fired (most of them are subsumed in the new company, L.G. Philips), because of a downturn in the economic cycle, the progress of technology and the success in streamlining operations. P. Zeven, the president of Philips Taiwan, does not admit they are giving up on Taiwan but hopes to transform along with Taiwan: 'As long as Taiwan keeps making progress, there is no reason for Philips to give up on Taiwan' (*Global Views Magazine*, 2001). PSK may have overcome its problems of manufacturing, raising its ability to produce and compete with other ATOs and making its position more secure. But the ongoing attraction of Mainland China, the setting up of the Suzhou plants and the ever-changing dynamic growth in semiconductor technology are upcoming challenges. Can PSK continue to grow and expand? Can it be the final beachhead for Philips in Taiwan?

REFERENCE

Global Views Magazine (2001), Cover story, Vol. 185, November.

9. Packing and testing in Taiwan's semiconductor industry: the United Test Centre Inc. (UTC)

Chia-wu Lin

In the semiconductor industry, the packing and testing of integrated circuits (IC) belongs to the second half of the production process. After the vertical disintegration system had been established in Taiwan in the 1980s, packing and testing foundries were also set up within Taiwan's Hsinchu Science-based Industrial Park (HSIP, see Chapter 3). UTC was founded in 1995, the trial run of machinery was held in February and production in quantity started in June (see Figure 9.1).

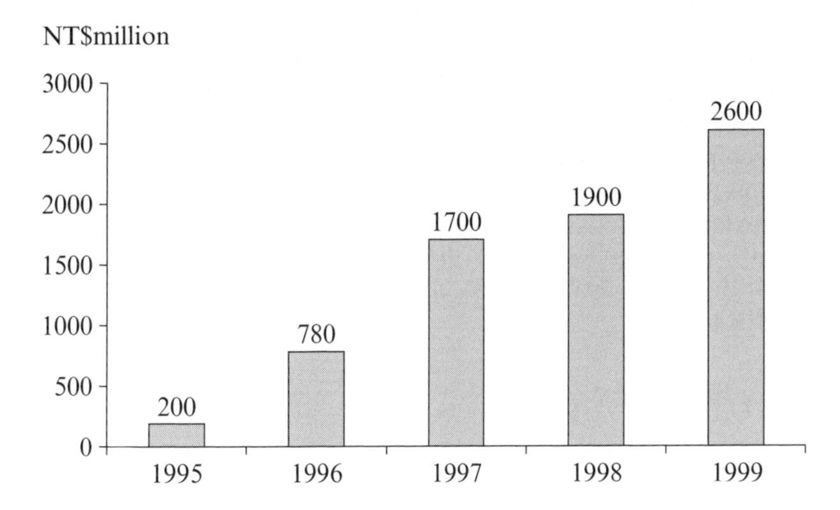

Source: UTC.

Figure 9.1 UTC: assets and turnover, 1995–1999, NT$ million

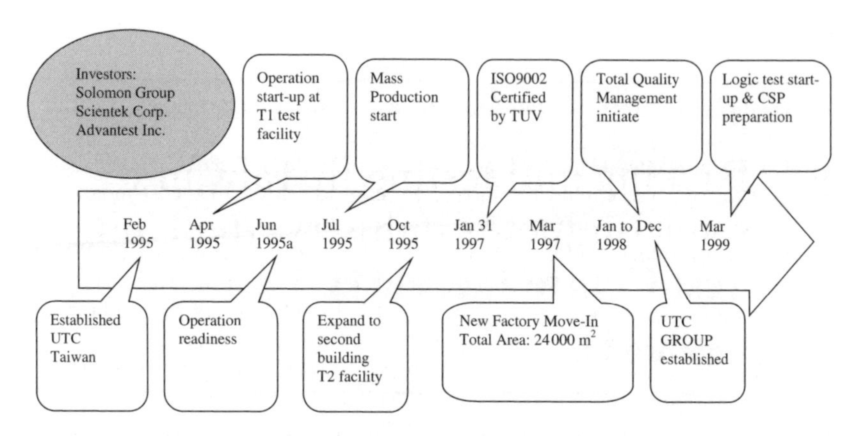

Figure 9.2 Milestones in UTC's progress

FOUNDATION AND PROGRESS OF UTC

The foundation of UTC was proposed by Japan's Advantest Inc. Advantest originally invited Vanguard to be a partner but Vanguard refused. UTC was finally based on the joint venture of the Solomon Group (30 per cent), the Scientek Corp. (30 per cent) and Advantest Inc. (40 per cent). The milestones in UTC's progress are shown in Figure 9.2.

UTC is a company providing IC and IC testing services, with some 830 employees, including 406 on-line operators and 275 engineers. UTC is a fully competitive semiconductor testing service provider, emphasising continuous improvement and engineer capability, on-time delivery, product quality, cost control and customer-oriented service. UTC highlights its quality policy, quality principles, corporate values and hopes for the future, by posting them on the company's bulletin board (see Table 9.1).

THE UTC GROUP

The so-called 'UTC Group' was established in 1998, with its main activities focused on long-term investments with the parent company or enterprises in strategic alliances. Advantest Inc. and three other companies (Singapore Technologies, Temasek Holdings and UOB) mutually built up UTAC (United Test & Assembly Centre Singapore). UTC also made an investment in Acer Test, another company providing testing services. The main shareholders of Acer Test are Acer, Chiao Tung Bank and UTC

Table 9.1 UTC's corporate values

- **Quality policy**: UTC is totally committed to quality products and services to our customers.
- **Quality principles**: (1) UTC expects to be a customer-preferred partner; (2) UTC hopes to be a reliable company. In order to achieve this, UTC uses 'Teacher' as its corporate culture.
- **Respect for values**: (1) Every employee is a partner of the company; (2) We are in an alliance with our customers; (3) The supplier is a business associate; (4) Continuous improvement is our way of life; (5) We accept the need for change as a reality; (6) Shared learning is the path to excellence.
- **Vision**: To be a distinctive advanced company regarded as such by customers, investors and employees. To be a forerunner in manufacturing with commitment to total excellence as a leading supplier worldwide.

Source: UTC.

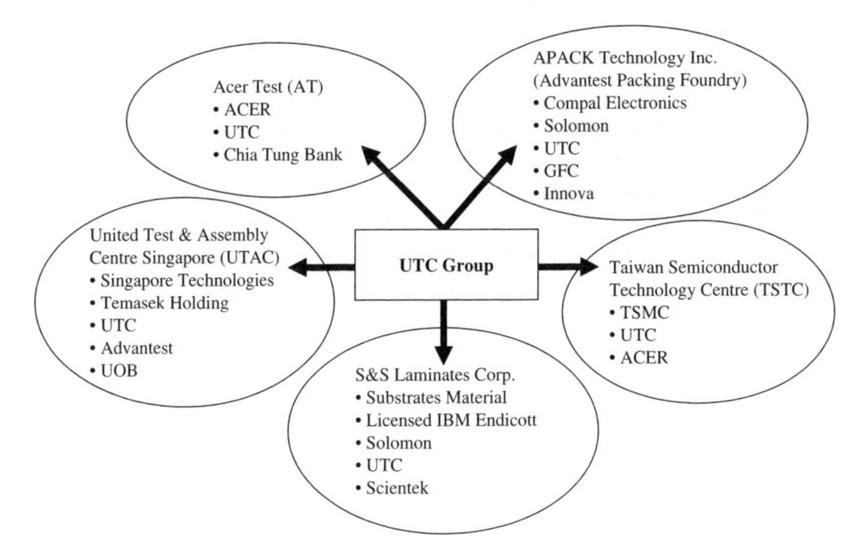

Figure 9.3 The UTC Group

(see Figure 9.3). The investments and the approach of mutual establishments are still mostly related to the second half of the IC packing/testing production process. The main consideration is to acquire different technologies and roadmaps of IC packing/testing.

UTC'S TESTING SERVICES

The testing service provided by UTC can be divided into wafer testing and final testing, the former providing 30 per cent of UTC's turnover. Final testing can be further divided into Memory testing and Logic testing, and on the current trend of product development, Logic IC will gradually take the place of Memory IC. Figure 9.4 is the flow chart of the wafer testing and Figure 9.5 is the flow chart of the IC final testing. UTC's technology roadmap (see Figure 9.6) shows that in recent years UTC has always followed the industry trend in transferring its technology step by step to provide high-level IC testing (Table 9.2 shows the development of memory IC).

UTC has also tried to enter the Logic IC market with high unit price and high profit and mixed signal process with a higher degree of difficulty in testing.

More importantly, UTC were also trying to expand their services to include packing as well as testing services. UTC planned to provide a testing service for Window BGA after the second quarter of 1999, and to provide a packing service for Window BGA after the first quarter of 2000.

STABLE AND RELIABLE TESTING QUALITY

Quality Organisation in UTC

In the semiconductor testing industry, the most important aspect after technology and capital is the provision of a *stable and reliable quality of testing*, in order to cement the satisfaction of industrial customers and maintain permanent partnerships. UTC emphasises the quality of testing service as part of its proclaimed values (see Table 9.2) and also within its organisation structure (see Figure 9.7). It set up a TQM office under the president and a Quality Reliability (QRE) division in the administration department, composed of three sections – Quality Assurance, Quality System and Quality Control. This division has 60 employees.

HOW TQM CAN IMPROVE TESTING QUALITY MANAGEMENT

UTC is in the first flight for hardware equipment, but the quality of service they provide does not satisfy its customers. UTC may therefore consider introducing total quality management (TQM) activity to improve the quality of testing services. TQM could improve six crucial areas (see Table 9.3).

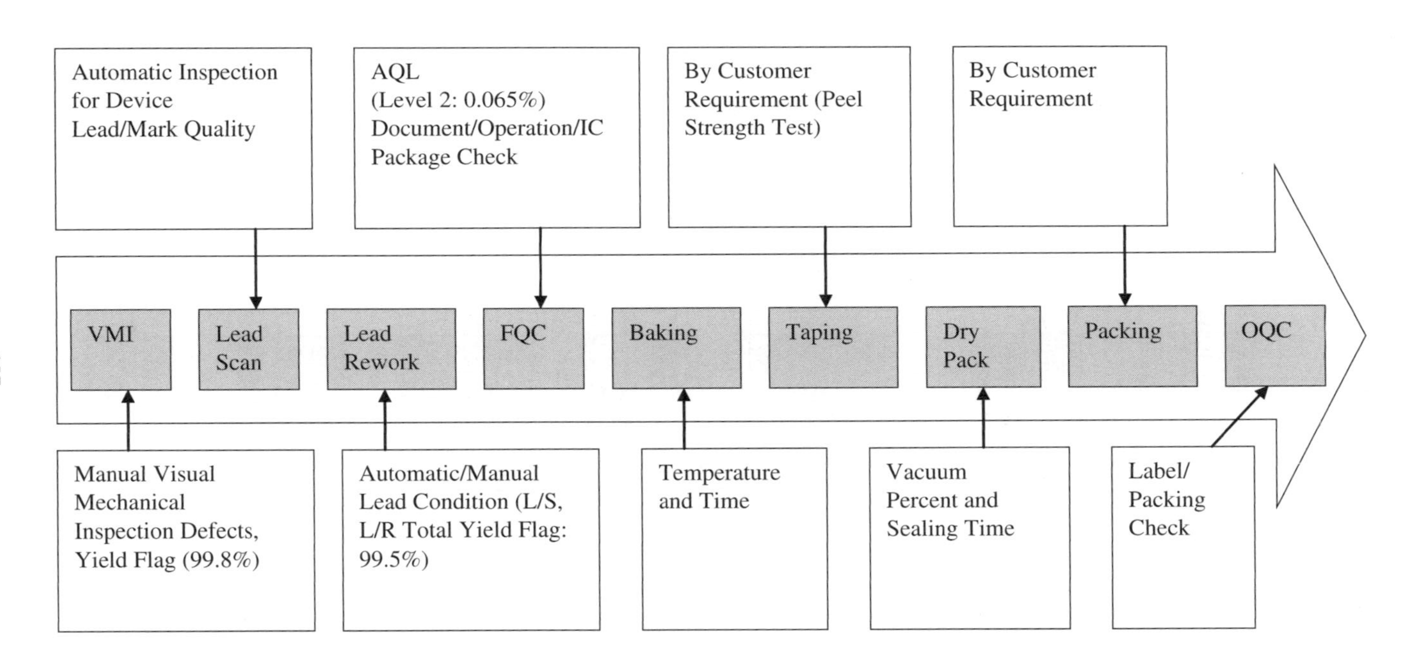

Figure 9.4 The wafer testing process flow organisation

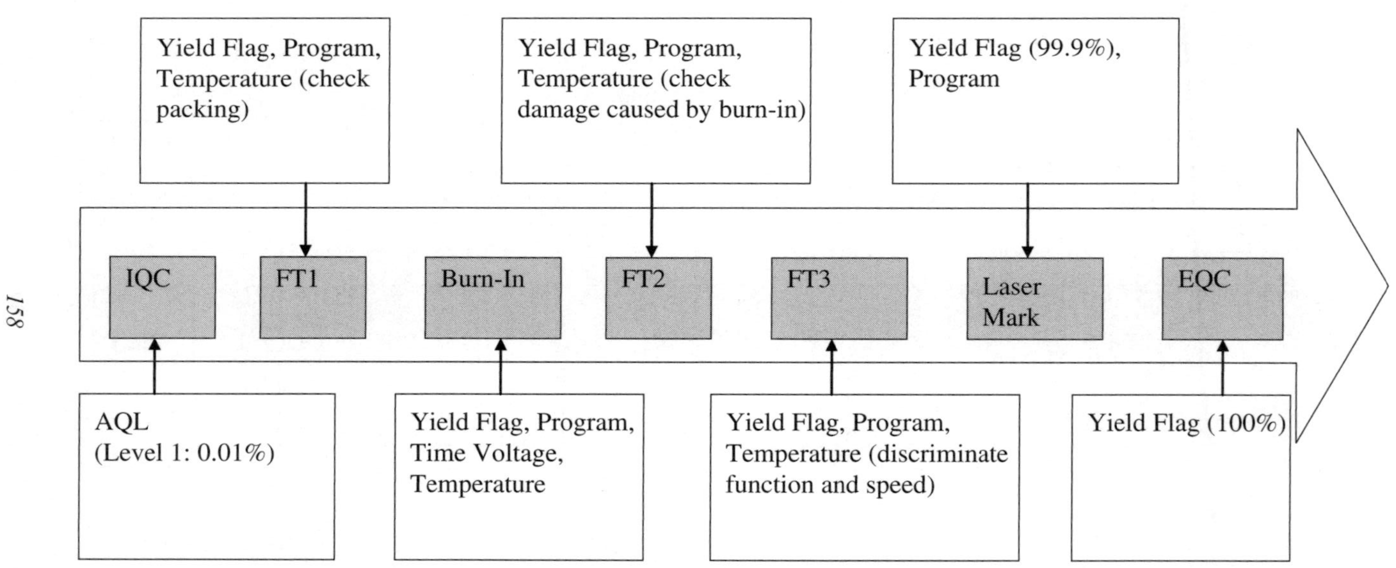

Figure 9.5 The final testing process flow organisation

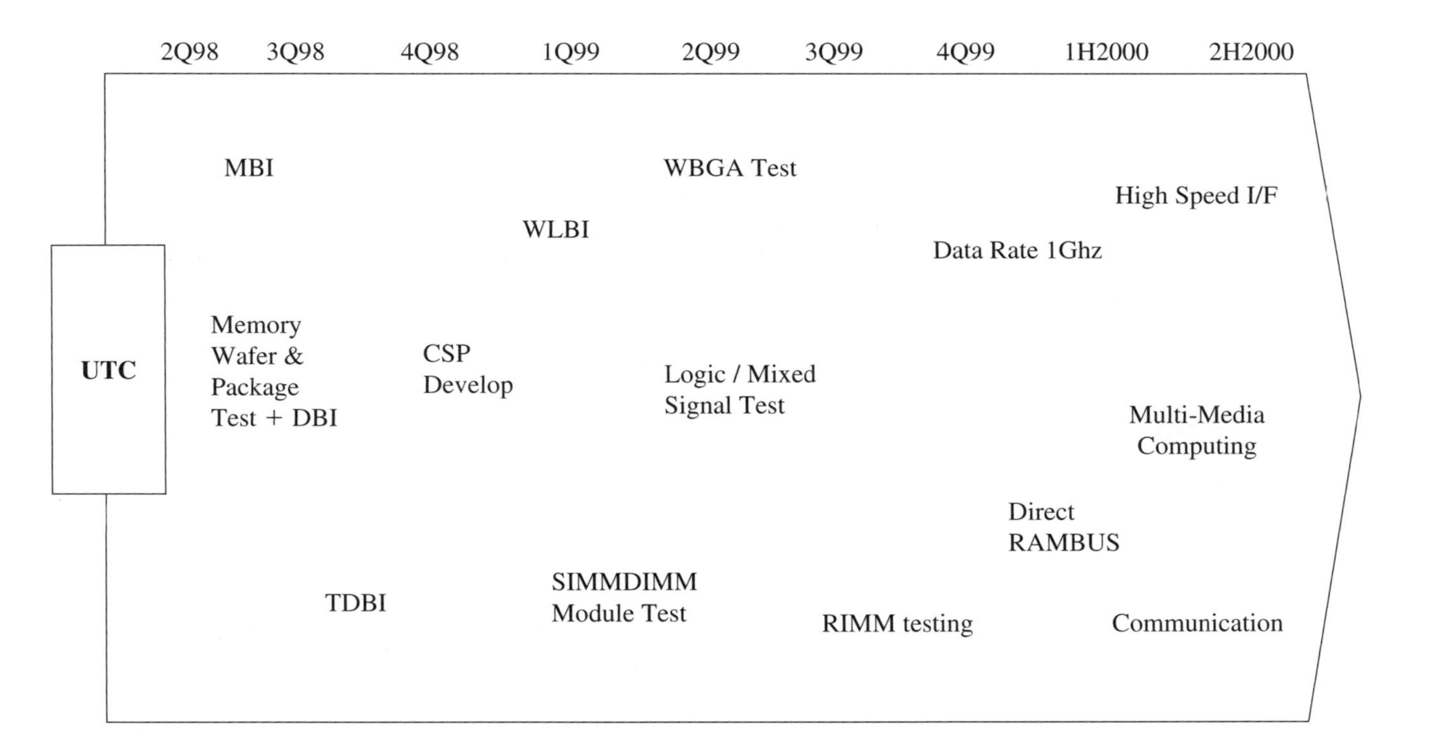

Figure 9.6 UTC's technology roadmap

Table 9.2 Development of memory IC, 1995–99

Year	Product	Memory (Mb)
1995	SSRAM	2
1996	SDRAM	128
1997	FLASH	16
1999	RAMBUS	256

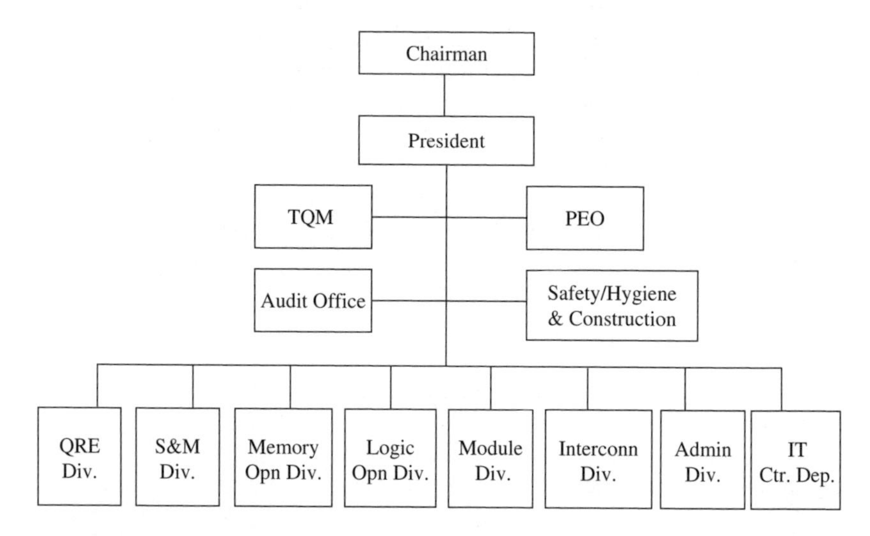

Figure 9.7 UTC organisation chart

Table 9.3 Key elements of TQM

- Follow-up policies that give customers satisfaction
- Inspire staff to participate in all aspects of test quality provision
- Confirm the quality that customers will accept, and adopt it
- Target the outstanding competitive factories
- Meet the competition on delivery time
- Establish a self-managed quality workgroup

UTC'S STRATEGIC APPROACH TO THE INDUSTRY AND TECHNOLOGY ENVIRONMENT

Characteristics of a Technology-intensive and Capital-intensive Industry

The semiconductor testing industry has to face continuous and rapid change in both technology and markets. The two major industry characteristics are its technology-intensive and capital-intensive nature. As the evolving speed of product development is so fast – from SRAM, DRAM and FLASH to the current RAMBUS – the upgrading speed of testing machinery must keep pace. Since each machine costs more than $NT10 million, the companies providing testing services have to face heavy costs to renew their equipment, and because the industry does not involve the processing and outputting of tangible products, the only charge that can be made is for services. With purchasing costs for machinery so high and the speed of testing is slower than that of the packing industry, packing must be at least around 30 per cent of the company's activities to make it possible for it to continue in business.

Mutual R&D for Technology

The president of UTC, Mr Tsai, feels that though the technology and specialisation of Taiwan's semiconductor processing have been fully developed, the R&D on more advanced technology is still far behind America, Japan and the countries of Europe. He suggests that all the companies should adopt mutual R&D on technology.

> Talking about semiconductors, they are very suitable for Taiwan since the logic had been set up. There are already many areas providing DRAM: Micron in the USA is one of them; in Korea, Hyundai and Samsung are the only two companies; in Europe, the only one is Siemens; in Japan, NEC (combined with Hitachi) is the only one . . . It's difficult for the companies in Taiwan to live unless these companies all merge together . . . Vanguard is the first one, then TSMC, and then UMC, etc. Acer might try this but it won't be the main force, is that right? And then, NANYA will be the next one. The point is how to link all the forces together instead of leaving them dispersed. If I give you NT$10 or 6 billion then maybe I can buy some technology. However, it's of no use that we just buy for ourselves. You can just buy the 0.175μ, but there are more models such as the 0.16 and 0.13. The later buyers will have to spend more money for more models . . . If all the companies co-develop together, and the only target is model 0.13μ, all companies can co-develop on model 0.13μ . . . and follow the professors in NCTU or NTHU for the development technology. There must be mutual technology and the money spent on it is necessary if the target is to be met. But Taiwan is probably different. It is difficult to promote cooperation for co-developing technology between Taiwan companies. (From an interview with President Tsai, 1999.)

THE PROBLEMS OF EARLY PREDOMINANCE

At the beginning of UTC's establishment, the main market demand was for high-speed SRAM, so the purchase of testing machinery was also mainly for SRAM. When the market demand for SRAM diminished sharply UTC suffered financial loss. Prices in the whole market were very unstable, especially in conditions of decreased market demand, and prices for packing/testing in IC processing naturally also fell sharply. In such a capital-intensive industry, the speed of capital recovery will be very slow in a depression: it may take as long as six years to earn back the capital, but some new machinery had to be developed during this period (see Table 9.4).

For testing on RAMBUS, with ten machines, 15 per cent needed high-speed machinery. The interview with Tsai shows how slow capital recovery and low price impacted on the fast progress of industry technology.

> Capital must be huge, and there is the quick advancement of DRAM product. The former speed was only 80MHz, and it remained unchanged for a long time for all machinery. In this situation, it would be OK if we made no advances for ten years. For the time on re-creation cycle, five years were also OK. The first time after the machines had been utilised, everyone was focused on speed, which affected many people and caused great disruption. For example, the speed of the 5365 we bought for the first time was 60 MHz. Then, in 1996, the request for high speed SRAM was proposed, and the speed became 100 MHz. Not long after came the speed of 250 MHz. So, if you had invested in the 100 MHz, you found that it was just a transitional medium. Consequently, without $NT3–5 billion, it's impossible for a testing house to support this kind of long-term development. (From interview with President Tsai, 1999.)

FIRST-MOVER OR FOLLOWER?

Under these conditions, being the first mover in the market is not necessarily a very good position. On the contrary, later movers can become dominant. They can avoid the cost (repurchasing equipment and machinery) coming from too rapid progress in science and technology and can

Table 9.4 Testing speeds, 1995–97 (MHz)

Year	Testing speed
1995	60
1996	100
1997	250

purchase some inexpensive used machines to provide a medium- to lower-level testing service; this gives them a big advantage on price owing to cheap costs. With these characteristics of the IC testing industry, a testing house business will have to seriously explore the question of the best time to enter during the industry development cycle. Should it be the stable period of mature technology development, or the starting period of new generation of technology? Furthermore, is it predicated on the progress and evolution of the whole semiconductor industry?

Academic students of the market life cycle usually advocate a company exploiting 'first-mover advantage'. Because first movers can set up the 'rules of the game' when the market is newly established, they can build up a more distinctive brand and get a larger market share. But this kind of advantage is not absolute, it can be replaced by new technology and a change in industrial structure. Is there really a 'first-mover advantage' in the Taiwan IC testing industry?

Tsai's answer is very cautious. He feels that what is necessary is not to make a choice at different stages, but to consider the roadmaps of machinery provided by equipment factories of different systems. These roadmaps are very important and can show the evolving trend of industry technology. After understanding the roadmaps of every system and machinery, the most important thing is to decide which stage of products and technology is the key market you want to be in. Then enter before the others.

> First of all, you have to understand the roadmap behind the whole machinery. For example, if Advantest is chosen, then you have to be very clear about the whole testing technique needed. While being clear about your aims, it may be necessary to be cautious. At the present time, if you'd rather give up market share and wait for the new products coming up, then focus your power on that and you'll win the whole world. If you take market share now, you'll have no competitiveness at all: no matter how much you take, you will lose it someday unless you follow up on price. If the later trade is very good and you're running a very good business, then you'll earn lots of money through entering the market at the right time. This is the question of choice. (From the interview with President Tsai, 1999.)

THE COMPETITIVE SITUATION OF THE TESTING MARKET

Unequal Relationship Between the Customer and the Factory

In this industry, the relationship between the service provider and the customer is not equal. The customer owns a higher level of technology on the writing of testing programs, and has more bargaining counters on requests for specification and equipment level of testing machinery because of industry competition. UTC found itself unable to adjust some product lines or equipment to respond to the depressed condition of low testing prices resulting from the low price of SRAM and DRAM for cost considerations.

> In the company's operation, service quality is an essential factor. Where the capacity of all Taiwan's machines is overloaded, you'll be willing to pay a higher price to get machines. On the other hand, if the capacity is still not yet overloaded and service quality is fine, then you may have priority to get machines but without a good price because the prices are all the same! Since the testing house itself does not write testing programs, which are provided by the machine vendors, once they have approved your company and put their production to the testing process, they depend on you for nothing else. What you need to do is give them a good service record on testing, making the reject rate as low as possible, then have good logistics. Because you will not be relied on for any technical support, the clients will decide the testing service charge. In other words, the testing service price is decided on whether you have 'know-how' or not; if 'yes' you can charge high fees. (From interview with President Tsai, 1999.)

LACK OF STRATEGIC ALLIANCES

For UTC, although it is more dominant than other competitors in machinery equipment and systems, it has problems in the operation and management of production lines because of lack of strategic alliance suppliers in the same group. For example, MOSEL are the strategic alliance suppliers for ChipMOSE; ProMOS, ASE, TSMC and NAY YA Plastics Corporation (Electronic Materials Division) are the strategic alliance suppliers for ASE TEST; Vanguard is the strategic alliance supplier for First International Computer, Inc. (FIC); Silicon-Ware is the strategic alliance supplier for SiliWare; and Acer is the strategic alliance supplier for Acer Testing. UTC's production lines are unable to operate at full capacity and in conditions of low profit not being fully loaded or having to stop work are both wastages of cost.

> From UTC's viewpoint, [alliances with] companies such as IBM are preferable, but there are still some efficiency problems. It's because of the volume and diversity of customers and they do not need a single product. The single product with ProMOS will be one product as a whole if they go to ChipMOSE, and it will be very easy to produce with efficiency up to 75 per cent. There are various kinds of DRAM within the product, and all the requirements and programs of different factories are not the same. Even the licensed mark will be changed at any time because the logos of every company are different. So efficiency is always not good, about 50 per cent or 40–50 per cent. Over a 30-day period, if there are three days with material shortages, then the 10 per cent profit you need will not be available. Therefore, I just talked of 45 per cent or 50 per cent. Sometimes if you want to promote efficiency, I think it will be very difficult.

For NEC to come in at present, it will make up seven units. Then you'll choose some companies with more quantity and better prices to produce first. Supporting some better companies to produce will promote efficiency. Try as hard as you can to cut down the number of customers, keep it as not too many.
It is difficult to raise prices when all your production line or equipment is not at work. At current prices, we are not able to invest in new equipment unless the testing product is from the strategic alliance suppliers of the same group. (From interview with President Tsai, 1999.)

STRATEGIC ALLIANCES AND OUTSOURCING

Taiwan's semiconductor industry is famous worldwide for its IC foundries. Because of the specialisation in Taiwan's semiconductor industry, many large international factories are willing to make strategic alliance with Taiwan factories and entrust production manufacture to them. Why are those large international factories willing to transfer such technology? According to Quinn (1999), enterprises generally adopt four strategic principles for success (see Table 9.5).

ACQUISITIONS AND MERGERS

An acquisitions and merger (A&M) strategy has become very common since the early 1990s, and not only in the communication and transmission industry. There have also been many cross-industrial A&Ms in the information technology and electronics industry. In Taiwan, the competitive merger between UMC and TSMC (see Chapters 6 and 5) made a great impact. Does a merger, expanding market share and increasing the property and production capability of an enterprise, really assure improvement in turnover and profit margin? Will the semiconductor industry revert to a vertical integration system? As we have seen earlier in the book, the advantages

Table 9.5 Strategic principles of alliances and outsourcing

- After strategic alliances and outsourcing, enterprises can concentrate more on developing capacity which customers really need.
- Enterprises can make more continuous innovation to provide more performance and added-value than their competitors.
- Enterprises can keep more flexibility to cope with competitors' strategy modifications.
- By adding surplus value, cooperation with other companies creates a long-term investment.

Table 9.6 Production quantity in Taiwan's semiconductor industry, 1995–98 (billion)

Year	Quantity
1995	2.7
1997	10.6
1998	30.0

Source: ITRI's IT IS plan, 1998.

of vertical integration include reduction in cost and improvement in coordination. On the other hand, vertical integration can also cause an enterprise to lose the capability to adapt to its environment and innovation.

The Trend to Market Centralisation Resulting from Competition

The semiconductor testing industry in Taiwan is a fast-growing market with keen competition (see Table 9.6). However, there were 22 factories to share this production. An immediate effect of the fall of memory unit prices and the Asian financial crisis of 1997 was that the price of testing also fell. Though there was tremendous growth in demand, there was no obvious increase in the gross profits and operating profits (ITRI, 1998). The five most prominent companies providing IC testing services in 2000 in Taiwan's semiconductor industry were ASE TEST, FIC, UTC, SiliWare and Vate (*Commercial Times*, 13 January 2000), and their output value is some 70–80 per cent of the Taiwan testing industry (Semiconductor Industrial Information Network, 1999).

ASE TEST AND OTHER MERGERS

ASE TEST, which was in first position, is trying actively to acquire some key technology via mergers. In May 1999, it bought 70 per cent of the shares of ISE Labs in the US for US$98 million. ISE Labs is the largest US professional IC testing factory, providing IC detecting services. ASE TEST lays particular stress on testing operation for finished products after the IC packing process, so the product lines of the two companies are complementary. After the merger, ASE TEST substantially increased its professional testing capability, improved its competitiveness and created a complete packing/testing outcome. The parent company of ASE TEST, ASE, has also merged with Motorola-Chungli and the other main company providing both packing and testing, ChipMOSE, has merged with the testing department of Microchip. ASE can now obtain outsourcing orders from IDM Motorola, and ChipMOSE can acquire logic testing technology from Microchip orders (Semiconductor Industrial Information Network, 1999).

CENTRALISING TENDENCIES

Five testing foundries own the greater part of market share in Taiwan. How do these centralising tendencies affect foundries and customers? Industry centralisation concerns the size distribution of factories marketing some specific product or composite of products, and is a very interesting aspect of market structure. Because the centralisation of an industry will make a company's market share bigger, it plays a very important role in determining market power, that is market behaviour and performance. What reasons drive the centralisation of an industry? Academics have pointed to technology levels, entry barriers, competition and mergers, the stochastic growth of markets, and so on. Take the high level of centralisation of Taiwan's drink industry, for example; Gau (1991) thought it was created by entry barriers, including early predominance of distribution channels, preference resulting from advertisements and the expectation that early-predominant factories would fight back.

From the point of view of the Structure–Conduct–Performance (SCP) research framework, industry centralisation is seen as a pointer to industry structure and a most valuable variable. The presumption is that industry structure will influence factories' actions (strategies) and performance. However this traditional paradigm is not supported by all researchers; many are still unable to agree on the real relationship between market centralisation and industry profit. The SCP model has nevertheless shown a positive relationship between regional market centralisation and profit, at least.

JOINT VENTURES AND STRATEGIC ALLIANCES

Technology Transfer

Within the semiconductor production process, every main factory has made long-term investment in, or tried to merge with, other companies in the downstream or upstream production process. We have already noted the MOSEL–ProMOS JV to create ChipMOSE and the ASE–ASE TEST strategic alliance. The crucial point is the transfer of key technology of every product, the so-called 'Turn-key'. UTC's President Tsai unsuccessfully tried to make a JV with other factories to form a packing company, and also tried to acquire the technology of the large packing factory, Anam Semiconductor Inc. (ASI) in Korea. When ASE completed the merger with Motorola-Chungli and the packing factory Paju in Korea the scale surpassed the ASI in Korea to make the first large packing factory in the world.

> As a first step, what we thought was to set up a testing house. The other companies had Turn-key while we did not have it, and we would lose out to them. Therefore, we started trying to establish a packaging division, and used it to support the testing house. Then, we set up the TSTC, a joint venture with TSMC, ACER and Compaq. After that, we hoped not to fall behind ASE too much on technology, so we chose ASI because it's the number one in the world. We've paid an amount of money, and they will transfer all their technologies to us within seven years. (From interview with President Tsai, 1999.)

Order Patterns

In addition to technology transfer, why will most factories try to seek partners for strategic alliances in the testing industry? Another factor can be found in the analysis of the factories' pattern of order taking. The pattern of most factories can be divided into two types – one is to take orders on their own and the other is to take orders from the companies in which they have made a long-term investment or with whom they have a strategic alliance. These two patterns each occupy some 50 per cent of the total. So if the testing factory can make a strategic alliance with a large enterprise which is dominant in both technology and production quantity, this will be very valuable. (Semiconductor Industrial Information Network, 1999).

Choice of Partner

As President Tsai says, when choosing a partner for a strategic alliance, or the technology requirement from an equipment provider, you must make sure that you find the best because there is no room for companies in third position.

> The technology roadmap between partners is very important. You must choose the one in first or second position, there is no room for the third position. It is the same with equipment, what you have to do is to choose the number one in the industry or production process to cooperate with. In this way, your development will be better. The number two company's existence will be harder, and there will be no room at all for the number three. So if you cooperate with the number three, then you might as well give up because it will result in your closure. Why? Because in the current competitive climate, every company will want to purchase things from the number one, and use the prices of the number two to bargain with the number one.

For semiconductors, it will not be so common to set up a fab because each device in the fab will cost about US$5 or 6 million, and a fab will need ten or more devices at least. Therefore, though the price of the number one is more expensive, the number two has its own price. Of course, it's related to the roadmap. If the qualities are the same, then it will depend on prices. In order to make business, the number two will cut prices. Based on the prices cut by the number two, you can use it to bargain with the number one. And the number one will definitely compromise with you, or you will cooperate with the number two. (From interview with President Tsai, 1999.)

UTC'S FUTURE STRATEGIC APPROACH

Advances in Packaging

Tsai emphasises internationalisation when talking about the future, and his twin concerns are to advance in the packing industry and to keep up with the mainstream of the global semiconductor industry.

Talking of global technology, it's a must to control the wafer stage, it's necessary to start from the wafer, and then go to the packaging. There are two types of packaging, one is for DRAM and the other is for BGA. The current packaging for DRAM is using TSOP [Thin Small Outline Package], but it will definitely change to CSP [Chip Scale Package] after 1999. Why? Intel has defined the packaging of RAMBUS as using CSP. The only factory approved by Intel is Micron in the United States, and it's the first one.

With global strength, it's possible to go to the top level. The only rule is that you must have the ability to perform testing. So we have to define the specification of testing at 0.8 μm pitch. This is also the specification of Intel, and of Micron BGA. So we will follow the same specification in the future. If we make a small change on our testing machines, we can also reduce our prices. If production capability is full now, then every factory will raise their prices; but if it's not full, then the price for you will not be satisfactory. Hence, we have to break through these limiting factors, then a better price will be available. After about 18 months, we will cooperate with IBM, if IBM is willing to invest. This is our future from the viewpoint of the global position. In the future, we will have our own wafer, our own packaging. If you ask me about this now, I'll tell you we don't have them, but we're planning, and we will become famous and amaze the world at once whenever we make it. In Taiwan, we'll be the best, not the runner-up. (From interview with President Tsai, 1999.)

DOES THE FUTURE LIE WITH IBM?

In order to enter the packing industry successfully you have to find a large international factory willing to transfer their technology. For UTC, there seemed to be a dawn if they could cooperate with IBM. Though IBM does not have the ability to define the product specification and production process as UTC can, its technology for packing materials is far superior to that of other factories. Intel has defined future production process technology for RAMBUS to be CSP; but on material science, Intel is unable to better IBM. So, UTC's projected strategic approach is to obtain the transfer of technology from IBM, or at least follow the large factories on product trends. President Tsai thinks that only by doing this will their efforts not be in vain.

> Why couldn't NEC keep up? Because the material they used came from Japan. The whole world is using the VT [Video Terminal] from Japan as the substrate. The material we use is from IBM. Only IBM develops the material on their own, for their own use, and never sells out or transfers to others. The substrate material we use now we call SMAIS [Space Media for Amenity, Intelligence and Image Systems]. We've signed a contract with IBM (September 1997) and this material has been put into production since June 1998. They gave us the licence and we were approved by them, so our material can be utilised directly on the product.
>
> The later section of the contract is all material science-related. When the boss of IBM was signing it, he said 'Hey Tsai, Intel is very capable, isn't it? However, the later half of Intel's life will be dependent on IBM.' He was very confident of this.
>
> For the present Taiwan should learn strategy from IBM. We should follow an excellent company so we can see our way instead of going in the wrong direction. Under these conditions, we know we will be safe for the next 5 years. (From interview with President Tsai, 1999.)

Why is IBM willing to transfer its technology to Taiwan? In addition to the 'cluster effect' resulting from the prosperous development of Taiwan's semiconductor industry (see Chapter 3), Tsai thought that the other reason was that IBM is more concentrated on the technology of the more profitable, first half of the production process, and will then transfer the medium and lower production process to Taiwan.

> The IO [Input/Output] with higher than 300MHz must use a special kind of material, other kinds of material cannot sustain the high temperatures used by packaging foundries in Taiwan now. So IBM needs to concentrate on new material development and transfer the lower production process to Taiwan in order to reduce expenditure on overhead. I believe IBM will contract with Taiwan firms if we charge low prices (if, for example, IBM can get a 25 per cent commission). (From interview with President Tsai, 1999.)

UTC AND MARKET RESPONSE

Taiwan's HSIP has accumulated wide experience of technology and UTC's main managers and engineers are all recruited from the park. However even when many knowledgeable elites with abundant experience come together, they will not necessarily be able to form an enterprise with good operations and performance. In addition, the testing industry that UTC had entered is actually an industry situated between the service industry and manufacturing industry. The characteristics of this industry make the relationship between internal staff (especially sales, engineers and quality controllers) and external customers critical.

Though UTC occupies third position in the testing market, some engineers say that their capability is so good that they can take more orders when the market is fully loaded. Because UTC has been established for barely five years, all the machines are still new and computerised, which is not necessarily the case for other testing factories.

> The advantage of UTC is the computerisation of the whole factory (this is impossible for some companies). Its computerisation is so thorough that it's famous. (From an interview with a NANYA engineer, 1998.)

> The predominance of UTC is because its machines are so large that there is significant room for the development. (From an interview with UMC engineer, C, 1998.)

UTC'S INTERNAL PROBLEMS

Why can't UTC make a breakthrough on turnover and profit margins? Views from UTC's key account have analysed six key internal problems.

Turnover of Staff is Too High

The first staff of UTC to contact customers is sales. However the turnover of sales staff is so high that the person in charge in the testing factory will often have to confront the problem of adapting to new sales. Each new salesperson will not necessarily get acquainted with each factory's needs and technical requirement, and this will tend to cause frustration to customers. UTC has not set up a position of medium-level director (for the sales division, it is the vice-president), nor given sales very much power. Sales is therefore often unable to make a decision on the spot and always has to go back to the company to ask for instructions.

> Each time at a UTC meeting, there is no one present who can make the decisions. The members attending meetings are always changing. Maybe it's due to the turnover of staff, and no one knows who can make the decisions. (From an interview with UMC engineer, A, 1998.)

They have no technical succession. Once well trained, they will leave. (From an interview with UMC engineer, B, 1998.)

UTC's PC has a high change frequency, seven to eight PC were changed within three to four years. And each PC has its own style, so each one gives a different quality of service. This is not good for us.

The pool of sales staff is empty. There is a gap between what is assigned from the top and the bottom level. And the gap is so large. (From an interview with Winbond PE engineer, 1998.)

There are so many handling hierarchies without fully authorised power, so we have to go up to the top to find the boss. (From an interview with TMT [Touch Micro-system Technology] PC engineer, 1998.)

Cognition Discrepancy Between each Management Stratum

Because of different ideals or ambiguous communication, customers have an agreed impression that there is a large cognition discrepancy between management strata and subordinates, and this is an obstacle in solving problems.

There is a discrepant ideal between management strata and subordinates so quality is uneven. All the UTC engineers work hard, but management strata give consideration only to marketing, not quality. (From an interview with UMC engineer, B, 1998.)

The most critical issue is management in UTC. Sales often can't deliver the message of the management stratum correctly to customers because of unclear communication in UTC. (From interview with Winbond PE engineer, 1998.)

There is a fault in the middle level of the management stratum so that it can't deliver a correct message effectively between management strata and subordinates in UTC. The process can't work because subordinates work hard without appropriate back-up. (From interview with MXIC PC engineer, 1998.)

Slow Response to Customers' Questions

Testing foundries must have close association with the account's engineers when something goes wrong in the testing process, yield rate is too high, or something is irrational about testing feedback data. This association with the account is poor in the performance of UTC sales and engineers. Most customers complained specifically about response time, because speed is a key point of winning in the semiconductor industry.

UTC service is inefficient, we have to endure UTC's delayed response when there is a great deal of yield. Specifically in the long vacation, the service is terrible

> even though UTC sales and engineers have a job deputy. The job deputy system doesn't work because UTC's management stratum doesn't care.
>
> The MIS is OK, but it doesn't serve the purpose of increasing time efficiency. They have to combine many systems so that processes holding and production lines can't work. (From interview with UMC engineer, C, 1998.)

> As to UTC service, the charge is high but quality is not good. The performance of UTC is poor, the sales department specifically. It takes at least a day to respond to any customer's question.
>
> UTC can't respond and answer customers' questions efficiently. (From interview with Winbond PE engineer, 1998.)

> The key problem is the lack of back-up processes, so that UTC engineers can't deal immediately with all kinds of situations when the computer is down and can't work. (From interview with NANYA PE engineer, 1998.)

Efficiency of Testing System is Inferior and Recycle Time is Overlong

Although UTC is more computerised than other testing foundries, it still seems to find it hard to create an equilibrium between software and hardware. The UTC computer hardware system does not work efficiently. Most client engineers mentioned that UTC was always late on delivery and even caused customers to suffer great losses when there was a severe slump in prices.

> Efficiency is poor in UTC. Because of the internal system, UTC's yield is probably half of other foundries'. The capability for internal detection of error is shocking too. We really don't understand why efficiency is so poor. (From interview with Winbond PE engineer, 1998.)

> We don't care about price, but recycle time. We were dissatisfied with UTC last year, because UTC didn't supply sufficient machines, so they caused severe shortages, and IC marketing prices were depreciating from $4.50 to $2.00 in two weeks at that time. (Interview with UTC engineer, A, 1998.)

> There is a circle of time delay: UTC doesn't deliver their work on time. Because of the character of IC, it takes a specific time for testing. We are extremely dissatisfied with the delay: UTC doesn't keep to schedule and causes NANYA to suffer significant losses. (From interview with NANYA PC engineer, 1998.)

Lack of Horizontal Cooperation and Coordination

Customers distrust UTC's internal horizontal cooperation and coordination, and think that UTC doesn't have a spirit of teamwork. UTC engineers always deal alone with customers' complaints without any internal support, and it is vexatious that customers have to arrange coordination with UTC sales, testers or engineers by themselves. This also influences testing quality and delivery time.

> The horizontal coordination should be sufficient. Only the boss's order can make operators cooperate with engineers in UTC. (From interview with Winbond PE engineer, 1998.)

> The subordinates work without any back-up in UTC, but with great support in other testing foundries. With no support, UTC engineers are always too busy to remember customers' questions and respond quickly. (From interview with NANYA PE engineer, 1998.)

> There are some problems with horizontal communication in UTC. The bad horizontal communication and coordination causes serious delays in delivery. (From interview with Vangard PC engineer, 1998.)

> We have to make contact with the person in charge, engineers, or testers of UTC by ourselves when an emergency happens. (From interview with TMT PC engineer, 1998.)

> The UTC engineers feel helpless in driving quality because team-work does not operate. (From interview with TMT PC engineer, 1998.)

Lack of Initiative

What customers often condemn in UTC salespersons and engineers is lack of initiative. UTC never takes the initiative in informing customers about new equipment buying, or equipment change. Customers read data and detect something wrong about the testing product by themselves, then tell UTC engineers.

> Even UTC PC engineers don't know that UTC have changed equipment and plates. We hope that UTC will let us know information about equipment. For example, we entrusted new product to other testing foundries because we never know what new equipment UTC had bought.
>
> UTC made many modifications but seldom let us know. For example, the equipment was prepared for us originally, but it was used to test other companies' product later. UTC informed us of this situation when we wanted to use the equipment, and the situation had a bad effect on our plans. (From interview with Winbond PE engineer, 1998.)

> UTC engineers must take time to review programs, setting of machines, and monitors. It is not reasonable that customers always detect the testing problem when they go to UTC. (From interview with TMT PE engineer, 1998.)

We hope that UTC will take the initiative in informing us as soon as possible whatever problem they detect, and not deal with error only when the customer detects a problem and makes a complaint. (From interview with MXIC PC engineer, 1998.)

REFERENCES

Documents

Commercial Times (2000), 'For Ensuring Testing, VIA Intends to Cooperate with Vate', Finance and Economics page, 13 January.
ITRI (1998), IT IS Plan.
United Test Centre (1998), *Introduction*.
United Test Centre (1999), *Company Profile*.

Internet Data

Semiconductor Industrial Information Network (1999), 'Observation on Taiwan IC Testing Industry', itisdom.itri.org.tw

Interview Data (1998)

MXIC PC PE engineers.
NANYA PC PE engineers.
TMT PC PE engineers.
UTC President Tsai.
UMC PC PE engineers.
Vangard PC PE engineers.
Winbond PC PE engineers.

Company/Institutional Data

Acer Testing, Acer Testing Inc.
ASE (Advanced Semiconductor Engineering Inc.).
ASE TEST.
ChipMOSE, ChipMOSE Technologies, Inc.
FIC, First International Computer, Inc.
ITRI.
Microchip, Microchip Technology Taiwan.
MOSEL, MOSEL Vitelic, Inc.
Nanya, Nanya Technologies Corp.
NCTU (National Chiao Tung University).
NTHU (National Tsing Hua University).
ProMOS, ProMOS Technologies Corp.
Silicon-Ware, Silicon-Ware Precision Industries Co. Ltd.
SiliWare, SILICONWARE Corp.

Vanguard, Vanguard International Semiconductor Corp.
Vate.
VIA (Very Innovative Architecture).

Books and Journals

Gau, K.O. (1991), 'The entry strategy to high density market', unpublished Masters thesis of National Taiwan University, Department and Graduate Institute of Business Administration.

Quinn, J.B. (1999), 'Strategic Outsourcing: Leveraging Knowledge Capabilities', *Sloan Management Review*, **40**(4), 9–21.

10. The Acer Group's manufacturing decision: to enter China?

Soo-Hung Terence Tsai and Donna Everatt

THE ACER GROUP IN 1997

Sales and Products

The Acer Group is one of the world's largest PC and computer component manufacturers. Associated Acer companies include the world's third largest PC manufacturer, and Acer's mobile computers, network servers and personal computers are ranked in the world's top ten most popular brands in their respective product categories. Acer is the market leader in 13 countries around the world, and is ranked in the top five in more than 30 countries globally owing to the strength of its core business (see Table 10.1). On the distribution side, in 1997 it was the seventh largest PC brand in the world, the ninth largest in the United States and the number one in the Middle East, Latin America and Southeast Asia. The company has achieved top three status in many developing countries and its target is to achieve a top ten position in Europe, and top five status worldwide.

Acer also has a lucrative US$500 million operation in OEM manufacturing. With more than 23 000 employees, half of whom are located outside of Taiwan, and 120 enterprises in 44 countries supporting dealers and distributors in over 100 countries, it is a truly global organisation. Acer enjoys stable, long-standing relationships with some of the most powerful computer companies, including Intel and Microsoft, and can leverage its buying power for bulk discounts, secure supply and economies of scale.

ACER'S PHILOSOPHY

Acer's global mission statement is 'fresh technology enjoyed by everyone, everywhere' and the firm was widely regarded as a worldwide pioneer in delivering high-performance PCs at accessible prices. In 1997 Acer introduced the low-cost multimedia PC, the Acer Basic II, for just under

Table 10.1 Acer financial profile, 1995–1997, (US$ million)

For the year	1995		1996		1997	
	Combined	Excluding TI-Acer	Combined	Excluding TI-Acer	Combined	Excluding TI-Acer
Total Revenue	5825	5262	5893	5346	6509	6132
Revenue Growth %	80.9%	81.4%	1.2%	1.6%	10.5%	14.7%
Net Earning	413	163	188	150	89	262
Net Earning %	7.1%	3.1%	3.2%	2.8%	1.4%	4.3%
Total Equity	1450	939	2008	1321	2065	1638
ROE	38.4%	23.9%	10.9%	13.3%	4.4%	17.7%
Total Assets	3645	2340	4192	3156	4758	3608
ROA	14.4%	8.4%	4.8%	5.5%	2.0%	7.7%
Net Investment in Property, Plan and Equipment	963	284	1347	418	1470	616
Working Capital	767	758	996	995	875	974
No. of Stockholders	90 000	89 000	123 000	122 000	155 000	154 000
No. of Employees	15 352	13 947	16 778	15 272	22 948	21 307

Notes:
(a) Organisation internal data.
(b) Due to the drastic drop in the market price of DRAM during 1996–97, the Acer Group's financial results excluding TI-Acer operations are provided to reflect more accurately the economic status of non-DRAM Acer Group operations.

US$1000. Media and observers in the PC industry had expected a PC at this price to initiate the next generation of PCs to be introduced into the marketplace. This focus on price competition was restructuring the PC industry and PC firms worldwide were struggling to manage eroding margins by lowering costs. The Acer Basic II was initially introduced in Acer's domestic market, followed by Greater Asia, Mainland China, India, Russia and the United States. The standard configuration was equipped with an Intel Pentium processor, at least 16 MB of RAM and Microsoft Windows 95. In countries further along the technology adoption curve, the Acer Basic II also included an 8x CDROM drive and 33.6k fax/modem for fast connection to the Internet.

COMMITMENT TO THE FUTURE

Stan Shih, the founder and chairman of the Acer Group and widely regarded as a high-tech visionary, had a long-term vision to transform the Group into a global high-tech corporation. Though fully committed to aggressively pursuing ever-growing segments of the PC market, Acer also began to shift a sizable portion of its attention and resources to the '3E' market – education, entertainment and e-commerce. Newly created ventures in semiconductors, communications and consumer electronics were expected to play an integral role in Acer's strategic growth, capitalising on prior technological competencies while complementing the development of the existing PC business.

RE-ENGINEERING THE GROUP STRUCTURE

To create an organisational structure to support this vision and enhance global competitiveness the Acer group was re-engineered in 1998 (see Figure 10.1). The modified organisational structure resulted in the creation of several new corporate functions and business development teams.

This reorganisation was adopted to fortify Acer's overall competitiveness in the light of what industry analysts saw as a disintegration of the PC industry in 1998 – almost every product was based on an 'open standard', resulting in competition in market niches where companies were fighting for market segment leadership. Whereas PCs had historically been Acer's core business, the company was forced into developing strategic new business divisions. Therefore it continues to develop its technological expertise in components for mainstream PC systems as well as peripheral markets,

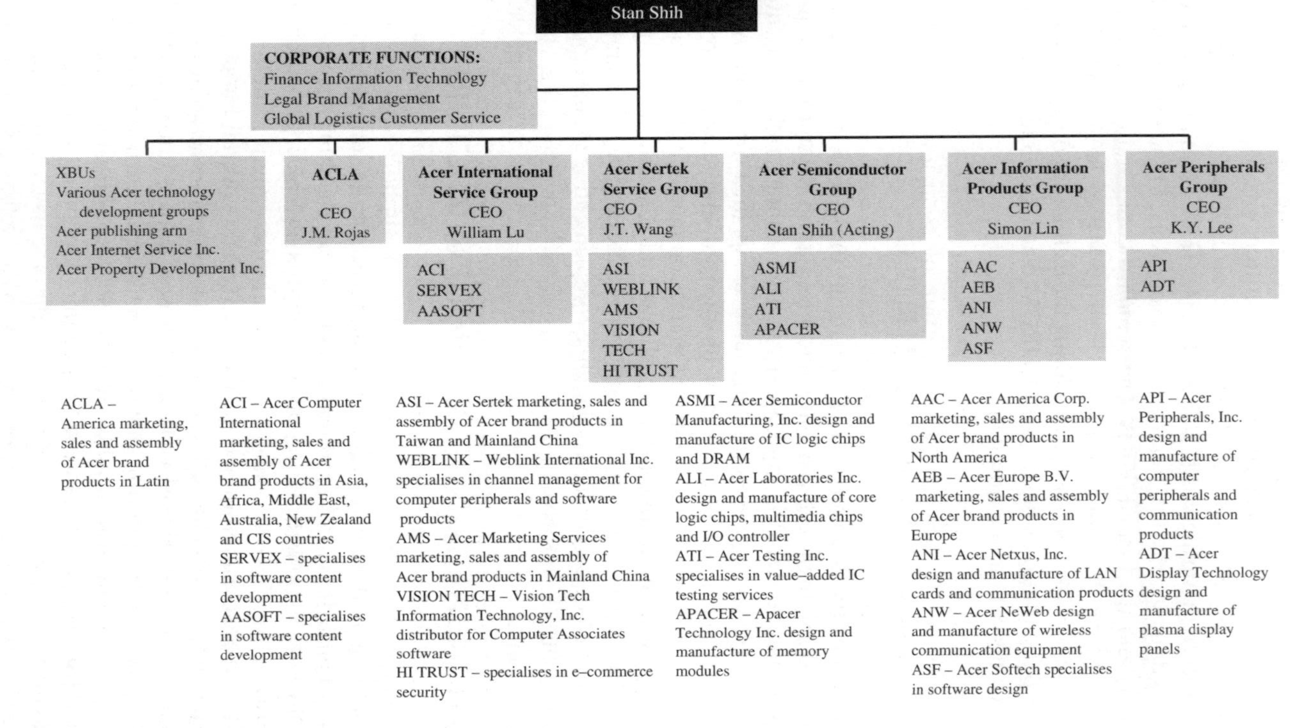

Figure 10.1 Acer Group organisation chart, 1997

but also seriously explores dynamic new opportunities in the consumer electronics, communication and semiconductor industries.

DECENTRALISATION WITH A CENTRALISED CORE

Centralisation of Corporate Functions

Each Acer Group member company operated independently, while working together to take full advantage of the resources available from a global US$8 billion multinational. The Acer brand name and the technological development brought about by Acer's R&D activities was the glue which bound the associated Acer Group of companies together, but it gradually became clear that centralisation of some common corporate functions would be a more effective method of managing the Group. In January 1998, four functions – Brand Management, Global Logistics, Customer Service and Information Technology Infrastructure – became controlled by Acer Group's global operations. These divisions were to provide the strategic direction, planning, integration and implementation of all related Acer businesses throughout the world.

Delegation of Responsibility

Acer's decentralised organisational structure delegated responsibility to management to involve employees in the decision-making process, a strategic advantage in the fast-moving, ever-changing world of computers. The associate vice-president of the Acer Institute of Education, Alan Chang, said that management at corporate HQ was willing to explain and justify policies to local managers, and was willing to 'take the time to convince the manager'.

Although there were several advantages in the autonomy of individual business units, senior management's biggest challenge was to consolidate the strength of the decentralised structure – to find a balance between Acer's core concept of a 'symbiotic common interest' which fostered personal commitment toward Acer's goals, and a second core concept of a highly decentralised, delegated management system that encouraged the head of each business unit to interpret and implement the corporate culture according to his or her own ideas, and achieve their specific mandates in the way he or she considered most effective. Shih believed that Acer would thus achieve a 'global vision with a local touch'.

Achieving a 'Local Touch'

Shih's 'local touch' was achieved in several ways. Acer's foreign market entry strategy involved forming joint ventures (JVs) to establish distribution systems and marketing/promotional activities, using partners who had intimate knowledge of the local market, and well-established relationships with local businesses and local distribution networks. Advertising and brand name support (marketing communications and promotion) costs were shared equally.

Product adaptation to suit local market language, tastes, trends, conditions and technological innovation was also carried out. Acer took an active attitude toward its local social responsibilities which helped them integrate into foreign cultures with a long-term perspective. Workers were hired locally and Acer gradually replaced Taiwanese managers with local managers, who were encouraged to be entrepreneurial and to feel like owners, and invited to participate in stock options, purchasing stock in their strategic business units (SBUs) at book value.

Another way that a 'local touch' was to be attained was by the creation of a global coalition of highly autonomous Acer companies owned predominantly by local investors and managed by local employees. Acer's strategy of floating various SBUs on stock exchanges throughout the world would not only permit institutional and retail investors to buy into Acer's success in a less intimidating way than purchasing shares in Taiwan (most investors did not follow the market), but also allow participation from local distributors who would have a built-in incentive to promote Acer computers to local buyers.

The 'Borderless Network of Companies'

This would be achieved with listings of a total of 21 Acer SBUs on stock exchanges throughout the world by 2100. This programme was referred to as '21 in 21'. The programme began in the summer of 1998, with the public listing of five Acer companies. The first listing outside of Taiwan where Acer Inc. was listed was in 1995 when Acer Computer International (which oversaw the distribution of Acer products throughout Asia, Africa, the Middle East and Russia) was floated on the Singapore Stock Exchange. Other companies in the Acer Group that were listed on international stock exchanges included Acer Peripherals, which manufactured colour monitors and keyboards, and Acer Sertek, a distributor of a full range of Acer products in Taiwan. Acer was intent on becoming a global player not only through international stock market listings but also through the global expansion of its manufacturing operations.

ACER'S GLOBAL MANUFACTURING STRATEGY

Sites, Assembly Plants and 'Uniload' Plants

The Acer Group in 1997 had 17 production sites and 30 assembly plants located in 24 countries around the world, which manufactured computers, peripherals and related high-tech components. In addition to the company's home base in the Hsinchu Science-based Industrial Park (HSIP, see Chapter 3) Acer had established production facilities in six countries (see Table 10.2). Additional production facilities were planned in North America, Europe and Latin America. Acer was one of the few IT companies in the world that had the manufacturing capabilities to produce complete product lines.

Acer's global manufacturing strategy involved not only expanding manufacturing plants throughout the world, but also shifting the assembly of computers from Taiwanese plants to areas where the computers would be distributed. These so-called 'uniload' plants assembled Acer-brand computers as well as computers for Acer's OEM customers (see Table 10.3). This business model (see Figure 10.2) ensured reduced inventory plus a faster time to market and was more responsive to changes in local market conditions. Being closer to the market meant that distribution logistics were more manageable and highly flexible. This approach not only increased responsiveness to local market conditions and trends but also reduced costs.

Table 10.2 Acer production facilities, 1997

- Penang, Malaysia
- El Paso, Texas
- Tilburg, Netherlands
- Subic Bay, Philippines
- Mexicali and Juarez, Mexico
- Cardiff, South Wales, UK

Table 10.3 Advantages of Acer's OEM business

- Uses any excess capacity
- Creates incremental revenue (US$500 million in 1997)
- Creates better understanding of modern design concepts, leading to improvement of R&D capability
- Increases power of OEM supplier in the client relationship
- Enhances global business relationship building

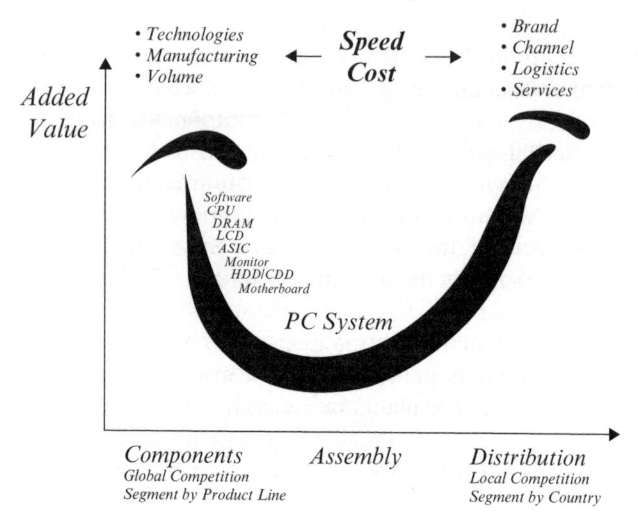

Figure 10.2 Smiling curve

Acer's 'Fast Food' Model

This approach involved moving the assembly of PCs to local sites, using components supplied by SBUs. Components themselves were referred to as 'perishable', determined by how much was at risk if the part was kept in inventory or not available. 'Perishability' was also used to describe how sensitive the component was to either changing technology or fluctuations in market price. The components were shipped via air freight to manufacturing sites worldwide to keep them 'fresh', and the most 'perishable' components (such as CPUs) were sourced locally. Non-perishable components, such as PC housings and power supplies, were shipped via sea because their 'freshness' was less of an issue.

'Central Kitchens'

Acer's 'central kitchens' were located in Asia and produced components such as motherboards, housings and monitors in Acer factories in Hsinchu, Taiwan and Subic Bay in the Philippines. Acer's globalisation strategy involved setting up 'central kitchens' closer to the world's fastest-growing markets and Acer was in the process of building manufacturing sites in

Juarez, Mexico, due to open before the end of 1998, to expand market share in NAFTA markets. These strategically located facilities were designed to help shorten the time from component production to final product delivery.

Customised Systems

Success on a global scale in manufacturing was due in part to the establishment of customised systems that transferred knowledge from R&D to manufacturing. Acer's manufacturing management had developed fully documented procedures for project management, similar to a project flow of critical events from the conception stage through development to the marketing of the final product, which facilitated the phasing out of various products in the marketplace. This system provided a step-by-step guide to phasing out inventory of parts and products, monitoring the number of spare parts required for servicing, and so on.

Local Control of Functions

The autonomy of overseas manufacturing operations was demonstrated by local control of many functions such as HRM and recruitment (under a set of general HQ guidelines) as well as all operational decisions, and ensuring compliance with local legal and administrative rules. Vendor contracts were centrally negotiated, but quantities and timing were decided by local managers to ensure their specific needs were met while capitalising on HQ's purchasing power. Vendor contracts were also secured at HQ so that Acer could safeguard its ISO 9000 Certification, which was partially dependent on the quality of inputs. In July 1997, Acer plants were also awarded ISO 14000 Certification which ensured that plants adhered to local environmental standards. Although the logistics to ensure global ISO compliance was a formidable task, Acer faced even more daunting challenges in the management of its global operations.

ACER'S BUSINESS MODEL

Shih's 'Smile Curve'

To explain the 'disintegration' trend, Shih created a chart that resembled a 'big smile' and called it his 'smiling curve' (see Figure 10.2). Value is added in the component production on the left side and marketing/distribution is on the right.

> The primary key to success in today's industry is providing value. By doing well in value-added business segments, companies can succeed in the current 'disintegrated' business environment.
>
> Today, there is no longer any value added in assembling computers – everyone can make a PC. To succeed in the new IT age, you have to gain a top position in component segments like software, CPUs, DRAM, ASICs, monitors, storage, etc. or else as a distribution leader in a country or region.

The key to success on the component side of the chart is global competitiveness. As Shih explains:

> Universal standards in components mean global competition, so if you're going to pursue a segment along the left side, you need technology and a strong manufacturing capability for economies of scale. On the distribution side, where competition is local, you can succeed through a good image, brand name awareness, well-managed channels, and effective logistics.

In the modern 'disintegrated' industry environment, there is 'one simple rule: if you are not one of a market segment's leaders, you cannot survive'.

> Whether on the right side or the left side of the curve, speed and cost are two main factors for success in such an environment. 'Speed' means fast time to market with new products and fast responses to change in the industry. 'Cost' includes minimizing overhead, inventory reduction and risk management. Only the leaders of each segment will survive, so whoever understands this curve will end up smiling in the future.

Many of the leading companies in the PC industry in 1998 concentrated primarily on marketing, and had little or no involvement in the component side of the business. Shih referred to such companies as 'computerless computer companies' and Acer was a critical link in their supply chain, giving Acer a high degree of power in its client relations.

Acer's Core Competencies

Client–server operations

The Acer Group organisational structure consists of Strategic Business Units (SBUs) and Regional Business Units (RBUs). SBUs are technology companies, responsible for the design, development and production of PC component and system products and also for OEM product sales and marketing. RBUs are primarily Acer-brand marketing companies, responsible for specific regional territories. RBUs develop new distribution channels, assemble finished products, provide support for dealer and distributor networks and create new JV Acer operations in key local markets. SBUs make Acer-brand component products and RBUs assemble and sell them.

Decentralised management structure

Acer's senior management believed that an IT company of its size could never be managed effectively from one central location because it would be impossible to react quickly enough to changing conditions in local markets. This was why responsibility was delegated to the SBU/RBU network of autonomous business units to ensure that Group companies competed efficiently in their respective market segments. The client–server structure let each unit benefit from the competitive advantages gained by leveraging the global resource base made possible by this large network of diversified business operations. Acer is a company designed to maximise its speed, cost and value advantages, responding quickly to changing needs and shifting trends. Acer minimises costs in order to maximise the value it provides to its customers.

GLOBAL INVENTORY MANAGEMENT

Inventory levels are predicated on market forecasts, which in the rapidly changing computer industry are difficult enough to predict accurately domestically, and far more complicated in international markets. Inventory levels that are too high create cash flow problems, downward pressure on prices and delay the launch of leading-edge products when older and obsolete products pile up in inventory; this can create a chain reaction of further difficulties, including a loss of market share and brand erosion. In 1997, inventory problems were a significant contributor to Acer's loss in the US market – US$75 million on sales of US$1 billion.

In his book, *Me-Too is not My Style* (1996), Shih used the following analogy to illustrate the challenges in multinational inventory management:

> This is like adjusting the water temperature when taking a shower. If the distance between the water faucet and the water heater is very long, it is more difficult to adjust the water temperature to a perfect level. Increasing the temperature may need twenty seconds, but one is already impatient after ten seconds so you adjust it further again. Twenty seconds later the water will be too hot, and one has to make the adjustment again. When the product supply is far away from the market, there will always be the problem of a time lag.

Being close to the market permitted significantly greater opportunities to adjust supply to market conditions; inventory turnover levels fell from 100 to 50 days after Acer implemented its 'fast food' model.

THE DECISION TO EXPLORE MAINLAND CHINA

Acer's thinking had been that securing a firm footing in niche markets was invaluable in gaining a footing in large markets – the 'surrounding the cities from the countryside' strategy which involved winning 'large cities' by overtaking numerous surrounding 'villages'. Acer's adoption of this strategy was predominantly due to the size of its 'home base': Taiwan's domestic market was no match for the enormous resources and size of other global players, such as the United States.

Development of Emerging Markets

Third-world markets represented the 'villages', and markets such as the United States and Japan were major 'cities' to be besieged through the storming of smaller 'villages'. Taking hold of these 'villages' or smaller territories would create a strong foundation on which to base Acer's assault on larger markets. Development of third-world markets would act as a catalyst to Acer's capabilities to compete in the developed markets.

Acer's 'Long Breath'

Another factor in Acer's decision whether or not to enter the Mainland Chinese market was the concept of 'long breath', a Chinese term referring to stamina and perseverance. For a company to maintain 'long breath', factors such as operational efficiency and high morale were crucial, but it was also contingent on giving up other activities to conserve energy for 'catching the big dragon'. Though highly diversified, Acer was careful to pursue only those businesses that were interrelated and developed competencies transferable to other operations.

Taiwan and China

Capital investment restrictions

Acer's competitive strategy was based on the development of various core competencies as a prerequisite to tackling its largest competitors on their home turf for a share of the US market – the largest market in the world, accounting for over a third of the sales of PCs globally. Acer was more familiar with marketing in third-world countries and had a more developed understanding of both the Chinese culture and the Chinese market, a prime consideration in its decision whether to pursue a higher profile there. Acer management was particularly concerned with the history of strained political relations between Taiwan and China and the risks Taiwan faced in investing in the Mainland. In 1998, Beijing was continuing to assert that Taiwan was not an independent state but rather a province of China. China's economic might gave it the power to sway countries in the West who, despite an ideology closer to Taiwan than China, saw immense and lucrative opportunities in the Chinese market. In Taiwan, however, the Economic Affairs Ministry advocated a slow and patient policy toward economic ties with China. Official policy restricted direct investment in the Mainland, and strongly discouraged large, high-profile investments by Taiwanese businesses there. Formosa Plastics, a Taiwanese firm which, despite restrictions placed upon it by the government and the risks associated with a large capital investment in China, constructed a US$3.2 billion power plant in China's Fujian province, only to have the Taiwanese government force it to sell the plant. Despite these risks, however, the Mainland market remained so attractive that Taiwanese firms had invested US$35 billion indirectly in China by 1998. Unlike the United Nations and the World Bank, the World Trade Organisation did not require statehood for membership, so Taiwan was seeking WTO membership. Complicating the issue was China's 'big China' mentality: Taiwan was dwarfed in size, economic power and political clout throughout the world by China and some Chinese felt that for this reason they could always get their way.

Transportation problems

Commercial ties with China were not only limited by Taiwanese government policy which discouraged investment, but also by restrictions on transport links across the East China Sea. It was only during the summer of 1998 that shipping links were re-established directly between Taiwan and the Mainland after 48 years of having to divert shipments south to Hong Kong before having them enter the Mainland. Although by mid-1998 Taiwanese regulations allowed Chinese ships to enter Taiwan's harbours, they still prohibited goods shipped directly from China to pass through their customs and prohibited Taiwanese goods from being shipped directly to China. This meant that while a shipping link was permitted, Taiwan could use it only for goods which were destined for international ports other than China. All flights from Taiwan to any city on the Mainland were diverted through Hong Kong, thus adding several hundred kilometres to travel between the two regions.

Sourcing inputs regionally

Although Acer did not face such high-profile political issues at other global manufacturing sites, there were issues the company would face in a Mainland manufacturing site that it had already experienced in the implementation of its global manufacturing strategy. An example of how Acer had faced another significant challenge, that of sourcing inputs regionally, could be found in Acer's experience at its Juarez, Mexico facility. The Juarez site was chosen because of its proximity to the huge US market as well as the fact that it was located in a *Maquiladora*, a special economic zone (SEZ) created by the Mexican government, which offered tax incentives, favourable business conditions and the infrastructure to attract multinationals to the region.

Originally, Acer had planned to import inputs from Taiwan for the Juarez plant; however it quickly became clear that this strategy would not allow the necessary degree of responsiveness and would make inventory management more complicated and less flexible. It would also add significantly to lead times in ordering parts. Acer, which in many cases was its suppliers' biggest customer, was able to persuade many of its vendors to join it in Mexico, an arrangement attractive to Acer's suppliers, owing to the volume of business Acer could guarantee and the fact that the arrangement allowed suppliers to manufacture for other customers. Acer's well-managed long-term vendor relationships meant that suppliers were considered 'family' and were willing to partner Acer throughout the world, contributing to Acer's, and in turn their own, success. Whether or not this practice could be applied in China was still unclear.

HR MANAGEMENT AT ACER'S MANUFACTURING SITES

Motivation for Working in Acer's Plant

Another major resource that would be sourced locally at the China plant was the workforce of several hundred people. It was expected that the HRM policy for workers at a Mainland China manufacturing facility would most closely resemble the profile of workers at Acer's Hsinchu factory in Taiwan, where almost a third of the workers were from another region (in this case, the Philippines), predominantly female, with an average age of 21 years. In China, workers migrated from urban communities to seek factory work in industrialised regions, commonly signing two-year contracts. Motivation for working in the plant would be based on the financial rewards such work offered. At NT$20000 per month in Taiwan (US$1 was equivalent to NT$33.2), a factory worker's salary was approximately four times that in the Philippines. The common practice in many large-scale Taiwanese factory operations of providing dormitory-style housing and meals, as well as transportation and basic medical care to workers from other regions, would be implemented in China as well. Acer could leverage its favourable reputation to attract the best workers, as employees from other regions would be assured of fair treatment and were guaranteed a reasonable standard of living and a competitive wage working with Acer. However it was expected that rapidly training the workforce to standard efficiency levels would generate significant challenges.

Shift Patterns and Overtime Pay

A highly disciplined and flexible workforce was critical for success at any of Acer's manufacturing locations. This meant that during peak production periods workers would have to commit to overtime, and when orders slowed, such as in the summer months, work light shifts. In Western manufacturing plants this was fairly easy to arrange with incentives such as overtime pay for workers, because they were familiar with adjusting for capacity in manufacturing environments. In China this degree of flexibility could not reasonably be assured, despite overtime pay, owing to the psychology of many Chinese workers.

The communist doctrine and the environment of the state-owned enterprise (SOE) had a strong influence on the attitudes of Chinese workers. A worker's pay in an SOE was guaranteed, regardless of the performance of the company or of the individual employee.The SOE system discouraged creativity and initiative, and indeed, showing these traits could create resentment and hostility among one's peers. Thus the underlying concept behind incentives or reward programmes was not fully understood by Chinese workers. Creating a disciplined workforce which was willing to 'go the extra mile' for Acer when required would be a significant challenge awaiting any HR and management team.

HR and Succession Problems

A critical success factor for the Mainland China factory would be Acer's ability to persuade highly skilled managerial and technical Taiwanese expatriates and their families to relocate to the site for a prolonged time. Many Taiwanese were hesitant to relocate themselves and their families to Mainland China and did not wish to take their children out of the school system in Taiwan: once children left the Taiwanese school system, it was impossible to integrate them back into it without being set back a year or two. A repatriated manager often also found that upon his return to Taiwan, positions of equal opportunity and status were difficult to find, and that being away had reduced his or her profile. Executives quickly perceived that an international posting was, more often than not, a career-limiting move, so repatriated executives were often recruited by competing companies to serve as their overseas business heads. These two factors resulted in not only a loss of valuable talent to Acer, but also a void in experienced overseas executives.

Despite the fact that Mandarin was spoken by both Taiwanese and Chinese, the two cultures were very different: many Taiwanese in fact considered their culture, background and experience to be more similar to the West than to that of China. This would mean great adjustment for both the manager and his family. Many amenities freely available in Taiwan were scarce in China so, in addition to adapting to a new and diverse culture away from their extended families and network of friends, employees and their families would also have to adjust to a lower standard of living. A critical part of the location strategy would be to pick a location that would be safe for expatriate workers and their families at a time of political and social unrest.

ACER'S STRATEGIC CONCLUSIONS

The economies of a move to the Mainland were nonetheless apparent. On the other hand, uncertainty regarding cross-Strait political relations created a risky business environment for Taiwanese companies investing heavily in China. Was the infrastructure sufficient to ensure effective logistics? What about the opportunity costs associated with such a large investment – was Acer missing out on more lucrative opportunities or were there alternative locations which would better suit its manufacturing needs? Critical factors on which the decision would be based, as we have seen, involved the difficulties involved in the transfer of the expatriate workforce, the disparity between the two cultures, as well as environmental certification concerns. The strongest rationale for manufacturing on the Mainland revolve around the economics of such a move (labour costs one-tenth of those in Taiwan). A manufacturing operation could also be a platform on which to expand Acer's presence in the huge market in China for either existing or new product lines.

Costs

A major input in Acer's manufacturing cost structure is wages. Labour costs on the Mainland are one-tenth of those in Taiwan, which would allow Acer to counter falling prices in the PC market. Acer's competitors have already made this move – thousands of Taiwan's PC-related manufacturers have invested in factories in China. Producing products with lower margins than PCs and peripherals, such as consumer electronics, is feasible in China though it would be attractive for Acer to capitalise on the lower costs associated with manufacturing in China to increase its profit margin on PCs as well.

Infrastructure

Is the infrastructure reliable in the Mainland China environment (see Table 10.4)? Can it reasonably assure on-time delivery of parts for production?

Materials Management

Can this function be managed effectively in China? Will the training of local supervisory staff be sufficient? Will the external uncontrollables (local suppliers' ability to remain constantly responsive, reliability of their delivery systems, infrastructure, and so on) cause losses? The infrastructure of industries supporting the production of PCs will also need to be analysed. Many inputs for PC manufacturing will be sourced locally, according to Acer's global manufacturing strategy. These inputs are required to meet very high quality standards to ensure meeting ISO quality requirements. Environmentally friendly manufacturing processes of suppliers, not well maintained in China, are required to meet ISO 14000 environmental manufacturing standards.

Safety of Expatriate Personnel

Acer senior management considered this to be one of the most critical of their decision criteria. To put their employees in danger, even when the economics of manufacturing in China were highly favourable, would go against the Taiwanese culture of treating their employees as 'family'. There is extensive literature that warns of the high failure rate of expatriate managers. The premature termination rate of assignments varied from 10 to 35 per cent, depending on the host country. And when there is significant cultural diversity, such as that between Taiwan and China, this figure could be as high as 90 per cent, based predominantly on the spouse's inability to adjust to the new environment (de Meyer and Mizushima, 1989).

Table 10.4 Infrastructure considerations

- **Energy supply**: Reliable source? Plentiful?
- **Environmental considerations**: Are there recycling facilities, for example?
- **Transportation**: Road, rail, port access, air services.
- **IT and technological infrastructure**: Reliable and up to date?

Despite the fact that the two cultures share the same language, there are many subtle and profound ideological differences between the Taiwanese and the Mainland Chinese. Management turnover will thus be a significant challenge and by using expat managers to run the manufacturing site in China, Acer risks prolonging the development of local managers, probably resulting in a decrease in their identification with the company and its culture.

Labour Laws

Labour laws in China are significantly less strict than in Taiwan, raising the question whether Acer's Taiwanese labour practices should be adhered to on the Mainland or whether it should adhere to local laws. Acer generally adheres to local laws – for example, visual quality control, which is repetitive and causes eye strain, is practised in Taiwan, though it is illegal in many industrialised countries.

Time to Realise Profits

If it takes a year or two before efficient production levels are reached, savings during that time will be undermined. Will the long-term benefits outweigh the short-term losses?

Political Factors

There will be risks associated with Acer's capital investment in China. Further deterioration of cross-Strait relations could result in expropriation, nationalisation or a government-dictated shut-down in operations. What is the degree of risk Acer faces in establishing manufacturing operations in China, and is that risk acceptable? How would such a move harm the company as a whole if it were to lose its investment, or a part of it, on the Mainland? What proportion of its assets would the China operation represent? Depending on the scale of operations, it would represent the capital costs of building, land, machinery, labour and inputs/inventory.

Acer would need to placate the Taiwanese government over their regulations and policies on investing on the Mainland. The Chinese government will possibly also require Acer to invest in other regional capital projects, or to guarantee workers' jobs. These costs must be factored into any analysis of the total costs in building manufacturing facilities on the Mainland. Manufacturing could be outsourced in an available Chinese factory, a viable short-term alternative that would allow Acer time to make a long-term decision based on the developing political landscape. Or Acer could purchase existing manufacturing facilities given the degree to which SOEs are being liquidated.

Financial Crisis and Liabilities

How resilient will Acer be in the face of a regional economic downturn? By 1999, there were signs of recovery throughout the region and Acer's financial management policies have followed a typically Taiwanese philosophy with regard to debt. The debt burdens which crippled so many Asian companies during the mid-1990s financial crisis were generally not present in Taiwanese companies. Taiwan's Finance Minister has said that the debt/equity ratio of the top 100 Korean industrial companies, for example, was five times that of similar companies in Taiwan. Taiwan's electronics industry in general bears a lower debt/equity burden than its US counterpart.

Managing HR in China

Managing a culturally diverse workforce is expected to be the most challenging aspect of the new venture. Can the Confusionist philosophy and culture found in Taiwan be integrated harmoniously into the diverse cultures found at Acer locations throughout the world (such as in China). Acer management believes that HR policy will have to be adapted to the Chinese environment and that the philosophy of a 'local way' will not be followed in the management of the Chinese manufacturing plants. Acer's management planned to adopt a more directive style with Chinese employees.

Technological Factors

Can a state-of-the-art manufacturing facility be built in China and will Acer's China manufacturing operations affect its ISO 9000 and ISO 14000 ratings? The key issues pertain to the incompatibility between (poor) local infrastructure and (advanced) technology.

The Zhongshan Operation

Acer finally chose Zhongshan as the location for a full-scale manufacturing operation (not a 'uniload' configuration centre). Construction at the site began in early 1998 and the 20 000m^2 plant was expected to be operational during the first quarter of 1999, producing Acer brand-name consumer products, as well as motherboards, housing and other components for Acer computers, eventually moving to desktop production. The plant's configuration allowed for additional capacity to be added. As had been the case in Juarez, many of Acer's major suppliers followed them to Zhongshan.

NOTE

For teaching and reference purposes, a companion case 'Acer Group's China Manufacturing Decision' (9A99M009) can be obtained from Ivey Publishing, Richard Ivey School of Business, University of Western Ontario, Canada. The authors are particularly grateful to the Jean and Richard Ivey Fund which provided support for the research fieldwork.

REFERENCES

De Meyer, A. and A. Mizushima (1989), 'Global R&D Management', *R&D Management*, **19**(2), 135–46.

Shih, S. (1996), *Me-Too is not my Style*, Taipei, Taiwan: Acer Foundation.

11. The Acer Group's R&D strategy: the China decision

Soo-Hung Terence Tsai and Donna Everatt

THE IMPORTANCE OF R&D TO ACER'S STRATEGY

The 'Aspire Park'

A major component of the vision for Acer promoted by Stan Shih, the Acer Group's founder and chairman, involved further development of the company's software expertise. Shih conceived an 'Aspire Park' which has been described as 'half Disneyland, half think-tank' to create Internet products and software, to improve Acer's competitiveness after 2000. Located about 30 minutes from Taipei, and the first of its kind to be developed by the private sector in Taiwan, it is a model of a high-tech 'island'. Involving an investment of US$8 billion over 10–15 years, the park was designed to be a full-scale residential community as well as an industrial park employing over 10 000 people, with 2000 households connected to the information superhighway through a high-speed advanced fibre-optic backbone, and millions of square metres of recreation, parkland and nature conservation zones – a luxury in the densely populated Taipei area.

One of the most important goals for the park was the development of Acer's software capabilities. The Creation and Innovation Centre aimed to develop Acer's software development skills and to contribute to the growth of Taiwan's capabilities. Aspire Park also encapsulated the concept of an 'academic village': people living, playing and working together could constantly exchange ideas, creating a hotbed for innovation and creativity highly conducive to R&D development.

ACER'S INNOVATION STRATEGY

Organisational Structure

Acer's innovation strategy could also be seen in their organisational structure, which was predicated on Confucian principles, including the basic belief in the good nature of man. Acer's senior management granted a high level of trust to regional and departmental managers, empowering them to make the majority of decisions regarding their operations. Empowered managers meant that the company was infused with flexibility and speed, two key success factors in the high-tech industry.

1998 Reorganisation

In 1998 Acer reorganised, reflecting the increased role of the company's software development capacities which had previously been a support function under the Information Products Group (IPG) and was now to be an independent business function (see Figure 11.1). R&D expertise and concentration of effort for the development of technology for PCs was very different from that required for technological development of notebooks and servers, and definitely different from that required for the development of software. Each division had its own marketing and after-sales support departments, allowing it to operate autonomously and with a more focused pursuit of its individual goals. Acer's two R&D labs, located in the United States and Taiwan, would play a role in developing the company's software expertise, as well as concentrating on other R&D projects to advance the strategic vision: the increasing importance of the role software development would play was the impetus for the reorganisation.

ACER'S TAIPEI R&D LAB

Commercial Development of Products

Acer's main R&D lab was established in Taipei in the late 1970s to conduct hardware R&D, and by the mid-1980s, had developed the capacity to conduct R&D for software products. The 700 managers and engineers at the lab focused on two major activities, the first of which was the commercial development of all of Acer's products and hardware components, which had generally been transferred from the United States as prototypes. This involved integration with hardware (in the case of software R&D) or other components, product testing, ergonomic design and customisation

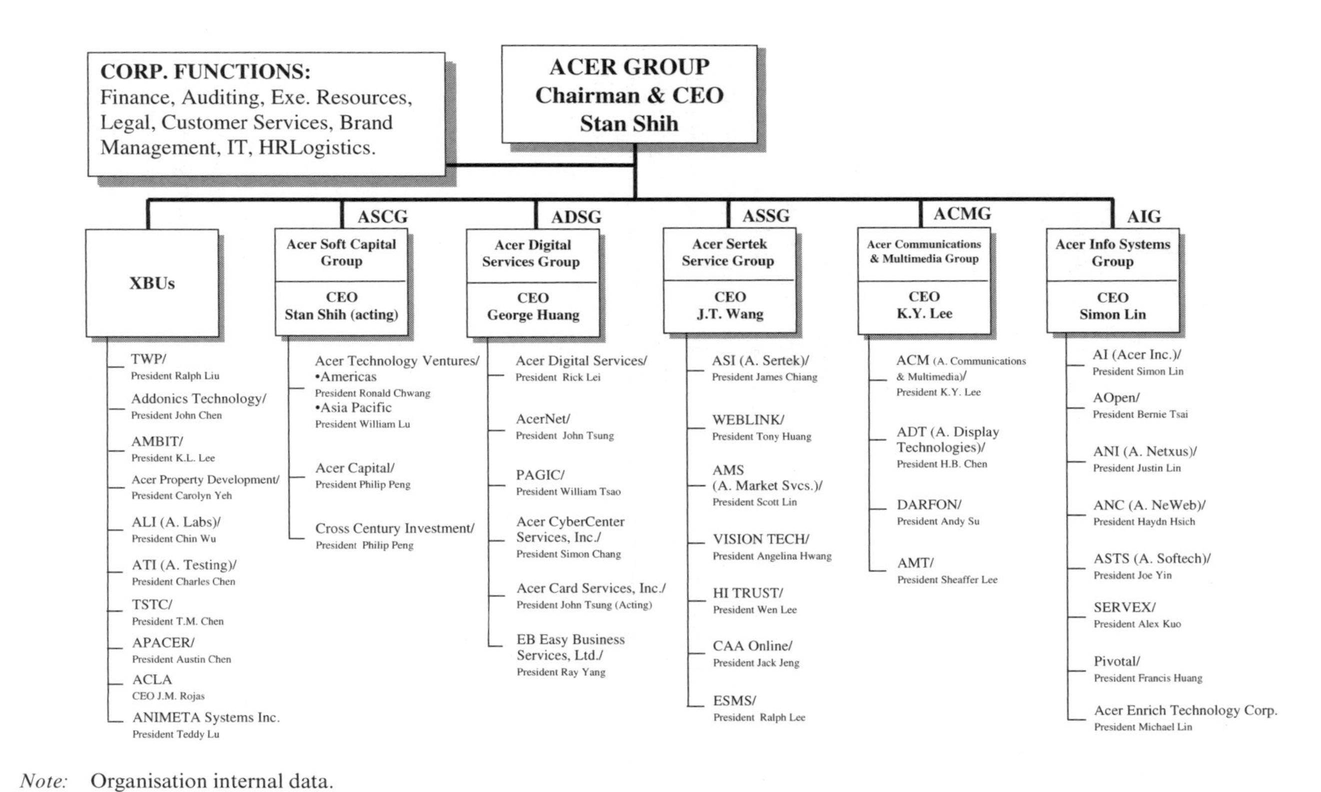

Note: Organisation internal data.

Figure 11.1 Acer Group organisation chart, 2000

for regional markets, based on input from regional managers as well as Acer's marketing divisions at HQ.

The second major responsibility of Acer's Taiwan lab was the commercialisation of electronic products for world markets, which required customisation for regional market 'clusters'. Though there were many similarities in product characteristics in international markets, customisation was required for countries at different stages of the technology adoption curve, as well as for regional preferences and local tastes, as we saw in Chapter 10. Acer's Basic II, which was initially introduced in the domestic market in 1998, followed by Greater Asia, Mainland China, India, Russia and the United States, had a standard configuration, but features varied between countries. An Intel Pentium processor and Microsoft Windows 95 would be included in all models but in countries further along the technology adoption curve the Acer Basic II would have more than the standard 16Mb of RAM, and also include an 8x CDROM drive and 33.6k fax/modem for fast connection to the Internet. Several versions of the Basic II and other technologies would thus be commercialised at the Taipei site, involving the recoding of the software required to run programs tailored to markets in Asia whose languages were not based on Roman alphabetical characters. Much of the world's software was originally designed in the West for languages based on the Roman alphabet and a few symbols. The '1s' and '0s' encoded in Western-developed software could not support characters in Asian-based languages and additional coding was required to accommodate Eastern characters.

Customisation of Software

These development activities were not only directed toward Acer's proprietary technologies and products, but also for those of other OEMs. R&D engineers at Acer's Taipei site developed expertise in customisation of software while concurrently generating revenue for Acer. As many Taiwanese OEMs were market leaders in their specific product categories in various countries throughout the world, a solid customer base for Acer's software R&D customisation activities was to be found within their domestic market. Many of these projects required a highly customised input and would not be transferable to any other parties. Revenue was generated through licence royalties as well as non-return engineering charges (NRCs), which involved customers pre-paying for work done on a project-by-project basis, and were non-refundable if the client chose to discontinue the project owing to a change of their business strategy or product focus.

Joint Ventures

To increase their competitiveness and access to knowledge, Acer also forged strategic joint ventures with other Taiwanese software developers to enter niche markets, such as writing software for OEM manufacturers and embedded systems (software installed in hardware that allowed control of all software programs) for hardware imported into the Taiwanese market. These and other activities at the Taipei R&D site were strategically planned with a theme of 'symbiotic common interest' with Acer's other R&D location in California's 'Silicon Valley'.

ACER'S R&D LAB IN THE UNITED STATES

Foundation

For several years the Taipei R&D site was the sole lab for Acer's R&D. However the need for a site located in the United States soon became evident, located closer to the pulse of technological innovation and able to develop a more intimate understanding of the global US market. Acer established an R&D site in the 'technology hub' of 'Silicon Valley' in the late 1980s; by 1998, the site had over 100 engineers, about 20 of whom concentrated solely on software R&D. Half of the engineers at the 'Silicon Valley' site were recruited directly from the United States, the other half transferred from Acer's Taipei site. As the lab developed, Acer's reputation in the Valley for innovative R&D attracted an ever-increasing percentage of local engineers. However the US labour market for computer hardware and software engineers was very tight, and many well-established labs with solid reputations were fighting for the best people among a small pool of talented and well-educated engineers.

JV Relationships

Acer's 'Silicon Valley' site focused on developing prototypes but the site was also responsible for locating many of Acer's strategic partners as an invaluable network of industry contacts. These relationships were formed partially with a view to establishing JVs with local partners, a strategy to which Acer was highly committed. Acer also created a venture capital arm to aggressively seek potential JV partners with a strong synergy with the company. This strategy would not be followed initially in China, however, since the country advantage lay not in the expertise of other software firms but in the abundance of highly trained software engineers (see below).

THE CHINA R&D DECISION

Workforce Advantages

Acer had had some exposure to business conditions in China after 1992 through JVs that established sales and marketing divisions, but had otherwise kept a low profile in the region owing to political sensitivities. By 1998, however, Acer was keen to establish a stronger foothold in China to position itself to capitalise on China's huge market potential. An R&D lab could be a precursor for a virtually inevitable deeper presence in China, including the expansion of manufacturing operations and marketing divisions to augment the existing sales and distribution network. However it was Shanghai's highly skilled workforce that was perhaps the most attractive of all potential reasons for locating an R&D site in the region. Universities in Shanghai turned out thousands of graduate- and doctoral-level technical students, among the best-educated in Asia.

The Shanghai R&D site, organised to operate as a wholly owned foreign enterprise, would commence operations with 200 engineers focused solely on the development of software. Given the abundance of highly educated engineers and the relatively inexpensive labour costs in China, Acer's initial research suggested that approximately nine software engineers could be hired in Shanghai for every one in Silicon Valley, and for every three in Taiwan. Acer would reach this pool of knowledge workers through Shanghai's universities, widely regarded as some of the top science and technical institutions in the world. The Chinese government was expected to benefit from the presence of labs such as Acer's, as the local engineers employed would acquire knowledge of systems and valuable R&D experience that could be transferred to indigenous Mainland technology firms. The Chinese government therefore took steps to ensure that the lab would largely be staffed with local employees. Acer would be assigned to partner two or three of most respected universities in Shanghai by the regional Chinese government, and hiring could be done exclusively through them, attracting the best and the brightest. (Restrictions on freedom of movement in China meant that Acer could hire software engineers only from Shanghai.)

Cross-Strait Relationships

Restrictions by the Taiwanese government meant that it would prove time-consuming and difficult to have any Chinese software engineers visit Taiwan. The policy of the Chinese government was to encourage knowledge-workers to relocate to China from Taiwan, thus increasing the transfer of knowledge to software firms on the Mainland. The reverse did not hold true, and the

Taiwanese government rarely granted visas and work permits for inhabitants of the Mainland. Since 1949, so-called 'cross-Strait relations' had been very tense, as Taiwan's pro-independence stance was very disturbing to Beijing.

COMMUNICATION CHALLENGES IN MANAGING INTERNATIONAL R&D OPERATIONS

Cross-cultural Problems

The fact that the Taipei and 'Silicon Valley' labs were on the other side of the globe from each other posed obvious operational constraints and logistical problems which, combined with cross-cultural issues, made managing communications between the sites a formidable task. The fact that engineers and managers from Taiwan were able to speak English helped, and through constant telephone and conference calls, email, Acer's intranet and videoconferencing, the engineers and managers at each site stayed in close contact. Acer also developed customised project management software which was shared electronically, so that engineers at each site were aware of the status of a project in its entirety at any point in time. In order to promote mutual understanding of each other's culture, as well as to transfer the knowledge and information for ongoing projects to Taiwan for commercialisation, US project leaders and the two chief software engineers from 'Silicon Valley' visited Taiwan for several weeks. Senior management from Taiwan also visited 'Silicon Valley' to acquaint themselves with operations as well as specific projects, such as the development of videoconferencing technology.

The Videoconferencing Project

Development of a prototype

The 'Silicon Valley' site had worked for several years on the development of a prototype for a videoconferencing project. To facilitate effective communication and a strong synergy between the engineers, the four main software engineers from Taiwan were transferred to the 'Silicon Valley' lab for six months. Though it added significant expense to the project, it enhanced the Taiwanese engineers' knowledge of the project and prepared them to act as the intermediaries for the transfer of the project's knowledge and expertise back to Acer in Taiwan. Though the communication flow was enhanced, there still remained ideological differences between the 'Silicon Valley' and Taiwanese engineers.

One of the key differences between the US and Taiwan labs was their perspective on how aggressively the latest generations of new technology should be introduced into the market. The US engineers were closer to the market and to the most recent technological innovations and industry trends, so it was logical that they should want to move consumers in every global market further along the technological adoption curve to exploit the growing power of PCs and peripherals. Though Acer's HQ was also eager to supply its customers with the best available technology, it felt it had a more realistic appraisal of the degree of technological innovation the market could bear. It was also concerned with developing technology that was compatible with its current manufacturing processes. The team in Taipei was more inclined to advance at a moderately slow pace on the commercialisation of the developments from the US lab than the engineers in 'Silicon Valley' would have wished. The US engineering team was working steadily toward the development of software that would accommodate up to 16 channels. However the specifications involved in this software made it difficult to use with the standard, existing channels for videoconferencing hardware. Management in Taiwan persuaded the US engineers to design software that could be bundled with Acer's hardware manufacturing base.

Compromise and cooperation

Many other issues considered contentious between the two sites were managed on the basis of compromise and cooperation. This environment of mutual respect was predicated upon Acer's culture, which fostered cooperation at every level.

Technology transfer

The engineers, though separated from their families for an extended period, found their visit to the United States to be highly valuable, recognising a wonderful opportunity to accelerate their careers by exposure to knowledge available in the US lab and its software engineers, who had vastly different educational backgrounds and experience, particularly as the lab was situated in a technology 'cluster'. They were able to experience first-hand the US market and culture, which would prove invaluable in their commercialisation efforts. The half-year transfer of these four engineers was successful on many levels and was used as a model for future technology transfers in large projects.

On a larger scale, Acer organised an annual 'Who am I?' series of conferences at Acer HQ, to facilitate the transfer of technical knowledge and certain confidential information between Acer's two R&D organisations. These conferences introduced Acer's various R&D divisions, project teams and acquired companies to other R&D divisions. As valuable relationships were discovered, informal meetings were arranged to discuss possible future alliances. Such meetings were invaluable in identifying possible means of collaborating with the other resources available throughout Acer's international R&D sites, as well as exploring the contributions of the JVs and firms acquired by Acer's venture capital arm.

ACER'S INTERNATIONAL INTELLECTUAL PROPERTY PROTECTION (IPP) STRATEGY

Though creating an environment that supported a high degree of cooperation was an imperative in successfully integrating international R&D sites, there remained other challenges if Acer's success in applying their global R&D strategy was to bear fruit. The development of an effective international IPP strategy was required to safeguard Acer's competitive position after 2000.

IPP Strategic Design

Acer's IPP strategy was designed to protect not only the company's traditional technological developments, such as hardware and software development, but also any proprietary business systems that senior managers perceived as core competencies, including such things as customised ways of servicing customers, developing HR or global logistic or inventory management systems. Acer's senior management also encouraged the application of the company philosophy of the belief in man's good nature, which they believed promoted honourable behaviour; however, various more concrete measures were also taken to ensure a secure IPP policy.

As IPP laws vary between countries and as the process requires ongoing monitoring, IPP is largely the province of local managers – not only is the manager more in touch with local laws, regulations and standard operating procedures, they also know their employees better, and can monitor and audit an individual employee's actions more closely and effectively than a centralised managing body. All Acer employees sign confidentiality agreements at the start of their employment, but the binding nature of contracts differs between countries, as does Acer's ability to litigate successfully if contractual infractions occur.

Prototype and Patent Protection

Measures taken to mitigate Acer's exposure to potential transgressions against their intellectual property code for hardware R&D involve Acer's legal counsel in Taiwan. They communicate on a regular basis with the managers developing any given prototype to apply related patents in the interim until the project takes its final form for commercialisation. This applies to R&D in the United States and Taiwan, and the patents provide a measure of protection in both countries. Once the product has been commercialised back in Taiwan, full patent protection is sought.

THE GLOBAL R&D DECISION

Would the development of a lab in Shanghai enhance Acer's global R&D strategy? Acer could leverage the country strengths in China, predominantly their abundance of highly skilled software engineers. On the other hand, there were some real obstacles to overcome before an R&D lab in Shanghai could contribute to the synergy between the lab in Taiwan and the United States. Logistical and management issues, as well as China's relationship with Taiwan, would pose significant challenges, and would require a sustained, long-term commitment. There were also potential IPP and cross-cultural issues. Should Acer make the commitment to a lab in China, or use the company's R&D resources to concentrate on building the technological expertise of the current labs?

Six Key Questions

The company faces six key questions (see Table 11.1).

The Strategic Importance of R&D

The globalisation of markets, combined with the development of R&D clusters throughout the world, makes it essential for Acer's enterprises and research institutions to be increasingly integrated into an international R&D network. A multinational's strategic plan to capitalise on new

Table 11.1 Key questions on R&D decentralisation

- How to best integrate the foreign sites with the culture of the home site?
- Should R&D activities be process- or market-oriented?
- Where on the scale of R&D development will the functions be placed?
- How far will the site's work be decentralised?
- How can knowledge transfer effectively be facilitated between sites?
- How can cross-cultural HRM be handled?

opportunities rests upon its commitment to, and the effectiveness of, its R&D function. Acer's success in branching out from mainstream PC systems and peripherals to new opportunities in software development, consumer electronics, communications and the semiconductor industry, will be largely dictated by its efforts and consequent successes on the R&D front. An integral factor in the success of any multinational's R&D function is the degree to which it is positioned to capitalise on strengths leveraged through an individual country's developing technology capabilities. Globalisation of the R&D function then becomes an integral part of the success of the entire corporation.

R&D 'Clusters'

R&D 'clusters' can be a part of a country's competitive advantage. Frost (1998) refers to a 'paradox of the trend toward globalization' where 'countries – and regions within countries – seem to matter more than ever in terms of creating world-class opportunities' (p. 144). As Frost explains:

> Globalization means that many high tech firms now find important centers of science and technology dispersed throughout the world, often in regional pockets or 'clusters' such as Silicon Valley (computers and electronics), Milan (fashion and design), and Wall Street (finance). (Frost, 1998: 144)

Benefits and Costs

What will the benefits and costs of establishing an R&D site on the Mainland be to Acer, and how can they be best accomplished? In such cases Frost concludes that:

> '[B]est practice' is contingent. In other words, there is no one right answer to the question of whether a firm should build an overseas technological capability, or keep R&D concentrated at home. Industry conditions, the bargaining power of foreign governments, and of course, the capacity of the firm to coordinate and control a distributed network of R&D facilities all play a role. (Frost, 1998: 145)

Market Drivers

What are the key market and technology drivers in the software industry that are creating pressures for Acer to internationalise its R&D function? Frost looks at market drivers – the forces that provide incentives to customise products and services to meet local market conditions. Frost also uses the term 'market drivers' to 'capture pressures placed on foreign firms by host country governments to establish local R&D facilities for technology transfer' (1998: 145). He also states that 'heterogeneous customer needs' or a market with tremendous potential will 'create a pull' for technology companies to locate R&D near particular markets, such as the Chinese market (1998: 145).

Technological Environment

Frost also states that conditions in a firm's technological environment are an integral factor in the pressure the company may feel to internationalise its R&D operations. In industries where technology is changing very rapidly there is considerable pressure for companies such as Acer to locate R&D facilities close to centres of the most advanced scientific work. Chinese software engineers have acquired the expertise to develop software which can accommodate Asian characters, which supports the move to develop an R&D site in Shanghai.

International Technology Strategy

As Frost points out, understanding the major market and technology drivers is a key first step in formulating an international technology strategy, but it is only the first step. Many other considerations need to be taken into account, including the firm's business strategy, its resources and the capacity of its managers to deal with the organisational complexity associated with an international network of R&D facilities. Frost (1998) identifies four relevant key challenges in this regard (see Table 11.2).

As Frost explains:

> [K]nowledge originating in the regions does not travel seamlessly across borders . . . it remains to an important degree embedded in the dense web of inter-personal and inter-organizational relationships that typically characterize these centres. (Frost, 1998: 145)

Why Develop a Foreign R&D Site?

Acer will need to answer seven key questions before taking its decision (see Table 11.3). These forces can be summarised in three broad categories – access to markets, technologies, and resources.

Leveraging Country R&D Strengths

United States

Acer would face several challenges if it took on Microsoft or Intel in the development of many types of software, owing to the entrenched position of these large players in many product categories. Many US players possess

Table 11.2 Key technology management challenges

- **Competing for worldwide talent**
  This is perhaps the biggest technology management challenge with an increasingly tight labour market for highly skilled scientists and engineers.
- **Globally monitoring and accessing sources of innovation**
  1. To keep abreast of key market trends and leading-edge technologies.
  2. Michael Porter's analysis of home country conditions (its 'national diamond') in *The Competitive Advantage of Nations* (1990) shows how firms cannot expect the home country to be advantaged in all aspects of the innovation process. Firms need to build 'virtual diamonds' – linkages to other countries and regions where important sources of innovation (R&D clusters) reside.
- **Managing technical operations in emerging economies**
  1. Is a local R&D facility to be viewed simply as a part of the cost of doing business in a particular country (the cost of market access), or is it to be an important 'node' in the firm's evolving global R&D network?
  2. Key success factors of a minimalist localisation strategy (if the local R&D site is a cost of market access) include keeping down the number of expatriate staff, developing an effective IPP strategy and proactively managing the expectations of government officials concerned with technology transfer.
- **Building effective cross-border technical networks**
  Successful distributed innovation networks have made major investments in coordination systems – largely people, but also physical systems such as corporate databases, intranets and videoconferencing technologies.

Table 11.3 Key reasons to develop a foreign R&D site

- To facilitate technology transfer from the home country to local markets
- As a response to host country pressures, to encourage localisation of technological developments as a condition of market entry
- As a public relations move (with a view to making a positive impression on the local market and local government officials)
- To reduce costs
- To capitalise on local resources
- To speed up development time through parallel efforts at several sites
- To increase local market knowledge and sensitivity

unparalleled expertise, have deep pockets and access to financing, and have well-established R&D operations in the United States. Historically low rates of unemployment and the shortage of well-trained technical workers have made it difficult to recruit engineers in the United States. The best and the brightest have typically sought work at one of Acer's more prestigious or high-profile competitors in 'Silicon Valley', thus increasing the competition for talented R&D engineers.

Acer was nevertheless able to partially staff its 'Silicon Valley' site with talented engineers from the local market, and also transferred a number of engineers from Taiwan who were eager to work in the United States on leading-edge technology. The 'Silicon Valley' site was responsible for monitoring and accessing the critical new technological innovations affecting the computer industry. Prototyping for many of Acer's products was done there, since the site was close to the market and had the talent and network to understand it. This marketing and technical knowledge then flowed from the 'Silicon Valley' site to Taiwan.

Acer's JV activities with US technology firms also played a key role in the transfer of knowledge to the home country. The impetus behind seeking JVs in the United States was to uncover any potential on which to capitalise and to nurture the different strengths each partner brought to the deal. Acer's JV partner provided the product specs and marketing know-how and Acer provided the engineering talent, predominantly in Taiwan. This strategy was a key factor in Acer's overall R&D strategy. Using the experience garnered from working with its JV partner, Acer then could develop its own appreciation of product specs and marketing know-how.

The strategic importance of the US software market was so great that each major software company announced and introduced its software developments and enhancements there, regardless of the financial returns or cost of such a move. Acer senior management recognised that the development of its software brand on a global scale required a strong presence in the US market, and it was prepared to accept any losses for long-term strategic gain.

Taiwan

The Taiwan R&D site was an integration hub and the location of the development of software and hardware programming. Other activities included the development of product specifications, through know-how transferred from the United States. The Taiwan site had the required HR base, and could develop it through knowledge transfer from the United States. The next evolutionary step for the Taiwan R&D was the transfer of experience and knowledge by developing strategic relationships with US JVs and forging synergistic partnerships with Taiwanese hardware and/or software vendors.

China

An R&D site in Shanghai would fit strategically into Acer's global R&D strategy. China's country advantage lay in the competence of its software engineers, considered by some to be the best and the brightest in the world. Key activities in the Shanghai lab that would strategically complement Acer's global R&D activities included the revision of software coding for adaptation to various Acer products as well as OEM hardware producers. By developing software under contract, Acer could generate revenue while gaining valuable experience. Given the labour-intensive nature of this type of customisation, it would be more cost-effective and productive for this (and many other software development activities which were equally labour-intensive) to be gradually shifted to the Shanghai lab.

The Shanghai lab would initially focus on the customisation of software to accommodate Mandarin and other Southeast Asian regional characters. Software engineers educated and trained in Asia had developed a distinct competence in customising, and the Shanghai site would prove instrumental in Acer's strategy to access and/or penetrate several Asian markets. English characters (which are one byte) are not compatible with Chinese characters (based on two bytes, or 'double-byte', technology which identifies Chinese and other Asian language characters). Since Chinese software engineers were developing this expertise it would be a major development area for the Shanghai lab.

An example of a typical Mainland R&D project would be Chinese language customisation for various software programs. Not only could Acer capitalise on the abilities of the Chinese engineers to develop software based on Eastern characters, but the engineers would also be more familiar with the interface, which would have to be designed to correspond to their market. Interfaces are related to a specific culture – a financial application for China would have its own accounting practices, system for issuing bills, invoice system and banking system – and all the software would need to be developed for the Chinese market.

Given the abundance of well-trained, less costly Chinese software engineers, Acer would also transfer labour-intensive activities to the Shanghai lab, including projects such as minor software coding enhancements, which is a highly technical and time-consuming process. This would give the lab team a chance to integrate itself into the Acer culture, while contributing synergistically to Acer's global R&D strategy. This work-for-hire strategy seemed very logical – product specifications would be defined by Taiwan or the US sites which had developed a competence in that area. This would leave the Chinese software engineers responsible for following directions from the project leader, whose primary mandate was to deliver software code for other Acer SBUs. Within one or two years some portions of software development currently done in Taiwan for the OEM clients could be transferred to Shanghai. A more flexible HR policy which encouraged creativity and initiative would then be required, but this policy could mature in tandem with employee development and growth in relationships between managers and employees at the site.

Challenges in Managing the Chinese Site

We can isolate six key issues here (see below). Three of these will require further examination.

- **IPP issues** would be protected by a strategy of managing projects in parts. Coding programs would be handed out in instalments to the Shanghai lab, with the source code being guarded in its entirety at the Taipei lab to protect Acer's intellectual property.
- Despite the fact that both the Taiwan and Shanghai engineers were of Chinese origin and spoke Mandarin, underlying **cultural differences** would impact on their ability to communicate effectively. Chinese engineers approached R&D from a different perspective to Taiwanese engineers and managers. Mainland Chinese had been socialised and educated in systems which encouraged strict application of orders, and prized rote learning over a display of initiative. Initiative was in fact viewed with suspicion and not positively reinforced in the educational system and in many workplaces. Taiwanese engineers showed initiative and creativity to a greater extent, and were ideologically more aligned with the US than the Shanghai engineers.
- Acer's culture of **empowering employees** could not realistically be imposed on the Shanghai lab. Acer's IPP strategy would have to be altered (as noted above), and the degree to which employees were empowered would be gradually phased in as the engineers became acclimatised to Acer's culture.
- Strained relations between Taiwan and China could at any time erupt into serious **political confrontation**, which would impact not only on the morale of the Chinese software engineers working for Acer in Shanghai, but on the viability of the entire operation.
- Restrictions on **transport links** across the East China Sea would affect not only the convenience (and therefore, perhaps, the frequency) of Acer's Shanghai managers visiting Taiwan, but also the knowledge transfer between R&D sites.
- There were several strategies in place to facilitate **knowledge transfer** to the Shanghai site and to develop the skills of the software engineers located there. Acer's primary approach during the first stage of development was to transfer internal knowledge through systems and technological expertise from sites in Taiwan and

'Silicon Valley'. The second stage would involve forging alliances with external software developers. Knowledge transfer could also be achieved through JVs with software companies in Taiwan and the United States arranged through Acer Capital, a venture capital firm.

IPP management

One of the biggest challenges to managing the Shanghai R&D site would be Acer's decentralised management structure, and above all its policy of transferring responsibility for IPP to the general managers at various sites. The prudence of such a policy in Shanghai needed to be examined, given China's reputation for IPP infringements. How could Acer reconcile its core organisational principle of trust in man's virtuous nature with an effective IPP strategy in China? Could the managers' ability to control and manage IPP issues in Shanghai be trusted? This was a critical issue because an effective IPP would play an integral role in deciding the future success of the lab. Acer's lead R&D site in Taipei assigned modules of source coding for development in China, which were then transferred back to Taiwan prior to the transfer of another module, thus preventing IPP infringements on the entire project. This safeguard would remain in place for at least two years, until Acer senior management agreed that the Chinese manager had reasonable control of the software engineers in China.

Infractions against Acer's proprietary systems and hardware R&D would be more difficult to control than infringements on software, which could be protected through IPP management software such as Microsoft's MS SourceSafe, and by controlling the source code in its entirety at the Taipei HQ.

Cross-cultural issues

This was expected to be the biggest challenge in the management of the Shanghai R&D lab. A different style of management would be required to ensure peak performance among Chinese software engineers, whose thinking and behavioural patterns were very different from those of the engineers in 'Silicon Valley' or at Acer's HSIP lab.

A great degree of success in R&D is attributable to the software engineers' ability to be innovative and creative. The Communist system meant that many Chinese workers were most comfortable with a passive style of management. Research into Chinese organisational behaviour lent some insight into common traits seen in Chinese employees (de Meyer and Mizushima, 1989): a high dependence on authority; participation by employees in discussion, but not decision-making; strict cultural norms regarding relationships; and avoidance of public conflict. Many of these characteristics run counter to Acer's culture and philosophies.

Chinese managers showed a high level of concern for the welfare of their employees, from both a career and a personal perspective. There was generally a high degree of security, perpetuated by the feeling that the employee was a member of the 'family', and would stay with it for their entire career. Though this pattern was not as strong in an R&D environment, evidence of it was apparent in the Chinese R&D lab. This potential for greater retention rates of highly skilled employees was a positive factor affecting the decision to develop the lab in Shanghai.

HR issues

Recruiting high-calibre personnel to staff the labs was increasingly challenging and costly. Once well-skilled engineers and technicians were found, another significant challenge was the relatively high turnover rates. There were also the problems in transferring management of the lab to a local manager, a process which could take several years and a significant investment of time and money for potential managers and engineers from Shanghai to come to appreciate Acer's style of doing business and, more broadly, Taiwanese culture as a whole. The relocation problems discussed in Chapter 10 would also be relevant here.

THE SHANGHAI LAB

Work at the Site

A suitable site in Mainland China was located in Shanghai. The R&D lab was created in October 1997 as an independent company, not under the umbrella of the Information Products Group; a departure from the existing organisational structure was consistent with the reorganisation of Acer's R&D structure as a whole.

For R&D projects conducted in Shanghai, Acer tapped into the vast amount of highly skilled software engineers available locally. Other Acer divisions or a JV partner provided the product specifications, market knowledge and know-how. Shanghai's revenue model was a work-for-hire approach, involving charging fees according to work performed on a project-by-project basis. The site took on many different types of projects to advance their software engineering expertise, and so was responsive to the varied software development needs of Acer's SBUs and the JV partnerships.

Management

Four expatriates were initially brought in to manage the site, overseeing 80 R&D software engineers, growing to over 300 before the end of 1998; between five and ten part-time Taiwanese managers split their time equally between Taiwan and Shanghai, with a mandate to increase the coordination of operations and to open communication channels between the two sites. There was also a succession plan to install local managers to replace the expatriates. In the interim, a team of locally hired managers acted as a bridge between Chinese and Taiwanese culture.

Lines-of-business and Later Reorganisation

In line with Acer senior management's recognition of the strategic role of software development capabilities, they then reorganised into new lines-of-business (LOB). Software development had previously been a support function under IPG, but was now to be an independent business function, since each product line was totally different and required its own focus. Later in 1998, Acer took reorganisation a step further and announced that the software division would be spun off as a separate company. According to Shih: 'Software is one of our key focuses in future development, so we are spinning off our software division in order to strengthen its development' (authors' own data).

THE FUTURE

In thirty years, Taiwan had developed into the offshore manufacturing centre for the US computer industry and the advent of the PC priced below US$1000 forced many US computer firms out of manufacturing to lower their costs and increase margins. This trend toward increased outsourcing among manufacturers such as IBM, Compaq and Dell contributed over US$10 billion to Taiwanese manufacturers such as Acer in 1998. The development of Taiwan's manufacturing industry fostered competence in activities further along the development chain, such as design and engineering and various synergies between Taiwan and the United States were becoming clearer. The creativity which produced innovative ideas and marketing acumen from the United States was a strong complement to Taiwan's manufacturing and emerging value-added technologies.

However the country's dream of moving even closer to true 'Silicon Valley' status seemed by the end of the 1990s to be in jeopardy. Industry observers argued that Taiwan's world-class production base could be duplicated too easily, and the high-tech competitive advantage it had developed was largely confined to the manufacturing realm. Taiwan was additionally at risk of losing a portion of its manufacturing base as operations increasingly moved offshore to less expensive countries, such as China, to capitalise on labour costs one-tenth of those in Taiwan. Taiwan's electronics information industry was vulnerable to price changes in the world market, as evidenced in 1996 when a worldwide collapse of semiconductor prices dramatically reduced revenue in Acer, as in many Taiwanese firms.

In order to ensure Taiwan's ongoing technological development, some industry experts felt it prudent for Taiwan to follow the US model of technological development. When the US lost its manufacturing base and competitive advantage in production technology to Southeast Asia, it turned to the development of advanced design microprocessors and chips to establish a competitive advantage. Acer's strategic move toward the development of its R&D capabilities was expected to parallel similar changes in Taiwan's high-tech industry.

However significant challenges lay ahead. Shih believes that the twenty-first century belongs to software development, not manufacturing in hardware industries. Taiwan's position with regard to software development is not competitive, and significant and immediate investment in the infrastructure of the industry and in development and training programs for HR development will be required – for both Acer and Taiwan – if they are to be successful in the next millennium.

NOTE

For teaching and reference purposes, a companion case 'Acer Group's R&D Strategy – The China Decision' (9A99M007) can be obtained from Ivey Publishing, Richard Ivey School of Business, University of Western Ontario, Canada. The authors are particularly grateful to the Jean and Richard Ivey Fund which provided support for the research fieldwork.

REFERENCES

De Meyer, O. and Mizushima, A. (1989), 'Global R&D Management', *R&D Management*, **19**(2), 135–46.

Frost, T.S. (1998), 'Building an international technology strategy', in M. Fleck (ed.), *Managing for Success*, Toronto: HarperCollins.

Porter, M. (1990), *The Competitive Advantage of Nations*, London: Macmillan.

12. 'Dragon flying high': carrying the legend to the new century

Soo-Hung Terence Tsai and Lena Croft

The previous chapters told us the stories of how Taiwan, a tiny island bounded by geographical constraints, emerged to play her dominant role in the global information industry. In this closing chapter we shall walk through these development paths and draw a conclusion as to the future of the information industry of Taiwan.

Our story started with Cheng introducing the background of the information industry of Taiwan. Huang then took the reader to the Industrial Technology Research Institute (ITRI), the cradle of technology, where highly qualified researchers worked around the clock to lay the foundation of the information industry in Taiwan. With the gradual maturity of ITRI, science-based industrial parks became the next strategic move in Taiwan to nurture high-tech personnel and innovation. Macronix International Co. Ltd. (MXIC), TSMC and United Microelectronics Corporation (UMC) are prototypes of entrepreneurship as a result of successful government polices to cultivate a favourable environment for business. Yet, invention and innovation needs well-developed equipment and services to realize ideas. Wu told us the story of Applied Materials Taiwan (AMT), partners of the value-chain. In addition to entrepreneurship, foreign investors also played their role. Philips Semiconductors Kaohsiung (PSK) is one of the early movers of the industry. The development path of the semiconductor industry in Taiwan is said to be different from that of the computer industry; Everatt has depicted an interesting story of Acer computers.

In the first chapter, Cheng offers a unique and systematic micro analysis to explicate the issue. First Cheng discusses that while the macro perspective argues for the interventionist roles of the government, market economy advocates posit the opposite. Planned economy advocates express an in-between position supporting selective governmental intervention as a critical success factor. For most, understanding the trends of the global information industry requires both the government and manufacturers contribute to the success of this little dragon. Yet an unqualified understanding of the global trend is insufficient. The implementation of different government policies toward two backbone information industries – the computer industry and the semiconductor industry – serves as driving impetus to lay the foundation for the distinctive achievement of the two industries of Taiwan. However the two industries exhibit somewhat different paths of development.

Cheng specifically identified these distinct paths and compared the salient features of success for these industries. The prosperity of the computer industry was a result of the government determination to eradicate arcade games manufacturing and learning through fabrication of local manufacturers. The semiconductor industry, on the contrary, is a prototype of governmental initiated and orchestrated product. The foresight of the government to develop Taiwan into a technological advanced information hub from a packaging centre heavily invested by foreign enterprises in the 1970s, permits the semiconductor industry to prosper. The differences in their development paths between the computer industry and the semiconductor industry challenge previous macro theoretical frameworks in explaining the success of Taiwan's information industry. As the semiconductor industry is presented as a government-nurtured success story, Cheng introduces the background history of the semiconductor industry and the rest of the chapters focus along the lines of the value-chain system of the industry. With the distinct infrastructure shaped by the government and industrialists as torchbearers, interorganisational networking production systems, together with entrepreneurship, allow the industries to excel. The conceptual framework of value chains is extended to examine the competitive advantage of the industry. By identifying the whole production process of the industry, Cheng examines how businesses and enterprises are involved in each link of the value chain.

Based on involvements of different parties to explain the success of the industry, this book is divided into three parts. Case analysis offers a solid base for discussion and understanding of how each party in the value-chain contributes its part. By adopting the concept of the 'smile curve', the development path of the semiconductor industry is depicted with the development and fostering of the industry being discussed first, integrated circuit (IC) design and manufacturing as the front end; and finally packaging/ testing and application of IC as the rear end of the 'smile curve'.

THE SEMICONDUCTOR INDUSTRY IN TAIWAN – THE ROLES OF THE GOVERNMENT

Aiming at nurturing Taiwan to become a high-tech industry hub in the 1970s, the government decided to establish a non-profit research institute to generate momentum for applied research specialisation so as to ultimately diffuse technology to small and medium size enterprises. The strategic policy was to merge three existing research centres under the Ministry of Economic Affairs (MOEA) including the Uni-industrial Research Centre, the Mineral Industrial Centre and the Metal Research Centre, and relocate the merged institute to Hsinchu, a county and city located in the northwest of Taiwan, about 80 kilometres from Taipei, the capital of Taiwan.

Once the strategic plan was approved, the government started to implement the policy with the goal of constructing 20–30 'intelligent' industrial parks by the late 1990s. Among these industrial parks or zones, Hsinchu Science-based Industrial Park (HSIP), under the jurisdiction of the National Science Council, was the only one that catered for high-tech development. With the support of the infrastructure, and skilled labour, the government offered a US$500 million subsidiary to the Science Park. Total land footage of about 1500 acres is well planned to cater for industrial, residential and recreational facilities as a complete service package. As for the software side, high-quality human resources provide support for expatriates and formulation of operation plans in addition to tax holidays, patent protection, financing and other incentives. The success of HSIP further encouraged the government to expand the strategic plan to form Tainan Science-based Industrial Park situated at the southern tip of Taiwan in 1995.

Other than offering land-use to realise this strategic policy, legislation was another move to support the whole plan. After the passage of legislation, ITRI was formed in 1973 consisting of three divisions and over 400 employees. Other than setting up policy to foster a research centre, a group of overseas Chinese, mainly from the United States, was invited to form the Technology Advisory Country (TAC) for organising technological cooperation as well as playing an advisory role. With the participation of specialists, technological transfer and licensing were made possible. These policies help to lay the foundation of the integrated circuit industry in Taiwan. Networking for both technology and manpower within Taiwan's high-tech industry was established through ITRI. Spin-off companies were gradually seen with the maturity of the industrial technology and manpower. The development of the industry comes to maturity when most businesses are established along the value-chain of the whole industry. However, the gradual maturity of ITRI with a significant increase in annual revenue to NT$17 million also indicated the time for reconsideration of the role of the government as well as the future direction of ITRI.

ENTREPRENEURSHIP – A VALUE-CHAIN APPROACH

With the government providing the infrastructure, such as the formation of ITRI in 1973 as a start-off base for the semiconductor industry, technology was transferred to small and medium size companies. Entrepreneurs were allowed to get involved in each link of the value chain. Their logistic support results in the smooth functioning of the value chain.

CASE STUDY: REALISATION OF ENTREPRENEURSHIP IN TAIWAN'S SEMICONDUCTOR INDUSTRY

The emergence of world-class semiconductor entrepreneurship in Taiwan is somewhat an extension of the cradle of advanced technology from the United States. If the role of the Taiwanese government is to foster favourable institutional environments for the development of science parks in Taiwan, then a few developed countries are seen as cradles for graduates from Taiwan to nurture expertise, to equip themselves with technological knowledge, strong social networks in developed countries, management skills, and then realise entrepreneurship in their motherland. Indeed, among these countries, the United States is considered as the prime nation for young graduates to explore knowledge and technology as well as western management. These strong linkages with the United States also permit the transformation of Taiwan into an 'American manufacturing backyard'.

Their entrepreneurship not only demonstrates their successful advancement in technology, take MXIC as an example, but also shares the vision and capability to fully utilise the financial market by creating a third category – technology stock – in the stock market of Taiwan. MXIC also gained access to the world's largest financial market – the United States – and successfully listed on the National Association of Securities Dealers Automated Quotation (NASDAQ).

THE INTERACTION

With all these entrepreneurs setting up their businesses along the value chains of the semiconductor industry, there is the chemistry that these interactions among MXIC, TSMC and UMC create a complementary and complete value chain to make a success story. Despite the success of the value chain, Acer, positioned at the end of the 'smile curve' and a case for an IC application, did not follow the traditional path of most investors from newly industrialised countries. Instead of entering developed countries to exploit the foreign markets, Acer imported microprocessor chips from the United States as early as 1977 when no governmental supports were received (Li, 2003). Their success story is somewhat different from those manufacturers along the value chain of the semiconductor industry. Re-engineering is a constant innovative programme within the company to ensure sustainability. With the technology acquired from leading information giants such as IBM and Texas Instruments, Acer is one of the few Taiwanese information technology manufacturers capable of developing its own brand name. Yet, observing the fierce competition in the United States, and lagging behind in technology in comparison, Acer has found its niche and expanded its business in Europe and Asia. Operations have lately intensified in the Mainland. Decentralisation policy allows individual operations in each country to make decisions and react to meet local needs.

POSITIONING OF TAIWAN IN THE COMPETITIVE GLOBAL SEMICONDUCTOR INDUSTRY – THE CHALLENGE AHEAD

Having gone through the development stages from a low-tech assembly centre for most leading firms from developed countries, the Taiwanese government sees the need to transform Taiwan to become a high-technology centre. The strategic plan to offer infrastructure and a start-off research and development centre for knowledge exploitation has successfully fostered its semiconductor industry. The achievements are profound. Yet any firms of global scale, or involving investments or alliances with other countries, cannot avoid competition with other countries which may follow the path of transformation of Taiwan and are offering cheaper labour costs to attract foreign investment. Competition across the Strait is seen as posing a threat to the future of Taiwan.

With China opening its door for foreign investment in late 1979, the nation has gradually evolved into a 'world factory' absorbing the lion's share of foreign direct investment from all over the world. On the lower end technological products, Taiwan faces tough competition from the Mainland. The accession of China into the World Trade Organisation has progressively removed most trade barriers. More multinationals eyeing the populous China market have shifted their production lines to the Mainland. Industrial parks in Shanghai, Beijing and Shenzhen are offering incentives, tax holidays, and an abundant supply of land and engineering talent (Wu and Chua, 2004). Furthermore, a rebate of most of the value-added tax (VAT) is given to Taiwanese investors on chips designed or manufactured in China (Spencer, 2004). On the higher end technological products, Taiwan is competing with the USA, Japan and South Korea. Among these leading countries, the USA is exploiting its technological edge to constantly come up with revolutionary products to change the face of the industry. Japan and South Korea are strong in applied research and have successfully entered the United States as well as the European Union and established their networks.

As for the United States, the unwillingness of the nation to provide research and development (R&D) support in contrast to European countries and East Asia, has dampened Taiwanese enthusiasm for the United States as the exploitation base for advanced technology. Spencer (2004) also raised concerns of a continuous reduction in annual funding of long-term R&D in microelectronics from about US$350 million in the 1990s to about US$55 million. In this respect, the European Union has substantially increased funding to support some regional programmes; for example, IMEC in Belgium has emerged to be an international centre for semiconductor manufacturing research.

Apart from these high- and low-end technological competitions, Taiwan also has to tackle advancement from the newly industrialised countries. Singapore recorded a 27 per cent increase in trade annually in 2001 and expected to go further when the import tax on most integrated circuits was reduced from 6 per cent to nil in 2002 in accordance with the WTO agreement (Wu and Chua, 2004). With its vicinity to the Mainland, Hong Kong has always taken the benefits of the non-direct trade across the Straits. With the gradual migration of manufacturing bases to the north, the Hong Kong government, though advocating positive non-intervention towards business sectors, has implemented certain policies such as the Closer Economic Partnership Agreements (CEPA), research grants to encourage collaboration between the private sector and local universities to keep up the competitive edge and to advance Hong Kong in becoming a high-technology hub. Lastly, the rising of India as a software engineering hub serving as an ideal outsourcing centre for jobs from the United States and other English speaking countries also poses new threats to Taiwan.

Like many others, Taiwan is facing the increasing competitiveness from those countries upon their advancement in technological development. The historical past of Taiwan might have permitted most graduates to be trained in the United States and practise their charismatic management style in their hometown. With the influx of students from the Mainland to the States, more and more talented graduates will be trained to upgrade the knowledge base in the Mainland in the coming future.

Globalisation drives all countries to join the international arena amid trade and foreign policies to negotiate and bargain for better terms. Future competition may be fiercer with the continuous fluctuation of prices for high-tech products, and for crude oil, a primary source for production. The increase in tension on petroleum supply for which Taiwan relies heavily on imports will put more burdens on manufacturers.

ENVISION THE FUTURE OF THE SEMICONDUCTOR INDUSTRY IN TAIWAN

While 'Vibrancy' is praised as a key as well as the answer to the success of Taiwan, the search for sustainable growth and competitive advantages continues to attract the interests of most people (Chang and Yu, 2002). To maintain its competitiveness, Taiwan will have to derive edges to ensure sustainability. For the role of the Taiwanese government, the continuous maintenance of a level playing field for entrepreneurs to compete is of paramount importance. Retreating from the active role of subsidising or orchestrating the development of the semiconductor industry by the end of the 1990s may seem to have avoided a conflict of interest in competing with business sectors. Yet the role of the government as a referee to ensure a level playing field is at least required. To that end, legislation to ensure the protection of intellectual property rights is not to be neglected. With the government as the upholder of rules and regulations, patent holders would effectively shape the norms and practices within the industry. This entry barrier forces other late entrants as well as other suppliers into a symbiotic relation with the patent holder in order to comply and follow. Isomorphism is observed within the industry.

Within the industry, a new product development strategy for IC design houses is advocated. Chen and Chang (2004) support the view that substitution of import products to meet the demand of local system manufacturers will be more successful in the semiconductor industry. Industrial clustering also allows manufacturers to focus on their own R&D and sales. In addition to the implementation of these strategies locally, international expansion is inevitable and is happening. Like other developed countries, there are concerns over gradual shifting of manufacturing lines to the Mainland. This move is considered as posing a threat to the future development of the Taiwanese semiconductor industry. However, with the easing of the restriction on semiconductor investments in April 2002, Wu and Chua (2004) see the possible potential of a complementary semiconductor industry structure between Mainland China and Taiwan.

The future relies on the continuous investment in R&D to create barriers for entrants and generate motivation for innovation, one of the biggest competitive advantages of the industry. The policy of the Taiwanese government to put in place measures keeping R&D, design, finance, logistics, and marketing – so called 'headquarters' functions – within the territories of Taiwan, is seen as a strategic move that will maintain the sustainable competitive edge of Taiwan. At the same time, Taiwanese investors could also join the arena to exploit the potential market in China, and continue to export their products to their US customers through the manufacturing bases on the Mainland (Spencer, 2004). Those who support the school of thought of international production networks would agree that most advanced countries would decentralise and de-concentrate their production lines in order to maintain the edge (Chang and Cheng, 2002). It seems like Taiwan is experiencing the same cycle. However there are projections that with the speed and capability of China in catching up, it will be able to develop the industry into a high-technology hub. It may be possible for the Mainland to replace the existing position of Taiwan which has fared better for these past two decades. Though many would disagree with this prediction, most would agree that Taiwan, by exploiting its existing well-developed technological base, will have to continue to focus on its R&D and become a high-technology information centre.

All in all, in order to survive in today's competitive environments, rising nations have adopted incentives to attract investment and technology in order to reduce the gap and compete with developed countries. Yet developed countries are also facing the challenges as to how to achieve a sustainable competitive edge and stay in the leading position. Those who are equipped with the capability and are constantly exploiting new knowledge and technology will certainly be the ones who can overcome the challenges ahead. The semiconductor industry of Taiwan is a vivid example to demonstrate the case.

REFERENCES

Chang, Chun-yen and Yu, Po-lung (2002), *Made by Taiwan: Booming in Information Technology Era*, Taipei: Reading Times.

Chang, Pei-chen Peggy and Cheng, Tun-jen (2002), 'The rise of the information technology industry in China: A formidable challenge to Taiwan's economy', *American Asian Review*, **20**(3), 125–74.

Chen, Chung-jen and Chang, Lien-sheng (2004), 'New product development strategy of IC design houses in Taiwan', *Journal of American Academy of Business*, September 5, 188–92.

Li, Ping Peter (2003), 'Toward a geocentric theory of multinational evolution: The implications from the Asian MNEs as latecomers', *Asia Pacific Journal of Management*, **20**(2), 217–42.

Spencer, William J. (2004), 'New challenges for US semiconductor industry', *Issues in Science and Technology*, **20**(2), 79–86.

Wu, Friedrich and Chua, Boon Loy (2004), 'Rapid rise of China's semiconductor industry: What are implications for Singapore?', *Thunderbird International Business Review*, **46**(2), 109–31.

Index